FOUNDATIONS OF THE UNITY OF SCIENCE: TOWARD AN INTERNATIONAL ENCYCLOPEDIA OF UNIFIED SCIENCE
Volume I in 10 parts and Volume II in 9 parts

Foundations
of
Biology

VOLUME I • NUMBER 9

Felix Mainx

The University of Chicago Press
Chicago and London

Foundations of the Unity of Science
Toward an International Encyclopedia of Unified Science

Editor-in-Chief   Otto Neurath
Associate Editors   Rudolf Carnap   Charles Morris

Committee of Organization

RUDOLF CARNAP	CHARLES MORRIS
PHILIPP FRANK	OTTO NEURATH
JOERGEN JOERGENSEN	LOUIS ROUGIER

Advisory Committee

NIELS BOHR	R. VON MISES
EGON BRUNSWIK	G. MANNOURY
J. CLAY	ERNEST NAGEL
JOHN DEWEY	ARNE NAESS
FEDERIGO ENRIQUES	HANS REICHENBACH
HERBERT FEIGL	ABEL REY
CLARK L. HULL	BERTRAND RUSSELL
WALDEMAR KAEMPFFERT	L. SUSAN STEBBING
VICTOR F. LENZEN	ALFRED TARSKI
JAN LUKASIEWICZ	EDWARD C. TOLMAN
WILLIAM M. MALISOFF	JOSEPH H. WOODGER

International Standard Book Number: 0-226-57584-5
Library of Congress Catalog Card Number: 75-131571

THE UNIVERSITY OF CHICAGO PRESS, CHICAGO 60637
The University of Chicago Press, London, Ltd.

Foundations of Biology
Felix Mainx

Contents:

Foundations of Biology

Felix Mainx

I. Introduction[1]

The word 'biology' is here to be understood in the sense of the fundamental science of living natural objects and thus as denoting all the disciplines of zoölogy, botany, physical anthropology, and the parts of neighboring sciences which are relevant to this field. Applied biology in its various branches, medicine among others, will be excluded. Human psychology, which, on account of the peculiarity of its special methods, requires a separate treatment, is also excluded. The notion of biology is not to be set up in opposition to the sciences of zoölogy, botany, etc., nor are these sciences to be subordinated to it, as sometimes happens (see Chap. III, B). Understood in this sense, biology is a branch of natural science, and it will be shown in the following pages that in its structure, in the kind of statements it contains, and in its mode of work, it represents an empirical science like the other sciences of nature.

The delimitation of biology according to its subject matter is a purely practical question. The concept of life was originally derived from the subjective awareness of human beings and then applied later to a greater and greater range of natural objects, when it was thought that these had properties in common with human behavior. In the present state of knowledge there is in practice no difficulty in distinguishing living objects from dead ones. Where there may be some doubt, as in the case of the viruses, it is purely a matter of convention whether, for scientific purposes, we classify such things under biology or not, just as drawing a boundary between zoölogy and botany is purely a matter of convention. Some authors attach great importance to

1. I am greatly indebted to Professor Joseph H. Woodger for his translation of this monograph from the German manuscript and for his criticism, and to Professor Victor Kraft for his reading of the manuscript and his valuable advice.

establishing the boundary between the living and the dead by definition and, from a metaphysical need, wish to treat this question as an ontological one (see Chap. III, B, 4). Even if we find that drawing a sharp boundary is not empirically possible, the independence of biology as a science will not disappear. Its independence rests on the observed peculiarity of its object and on the development of its own methods of research and points of view, which are required by that object. The division of biology into its various branches is also purely practical and for that reason depends, at a given epoch, on the prevailing direction of interest among the various possible points of view. Also for that reason, it is not meaningful to ascribe a fundamental importance to a "system" of biology. The vague separation into zoölogy and botany, according to the kind of object studied, is traditional and therefore still largely provides the basis for the organization of teaching and research. The subdivision into morphology, physiology, genetics, ecology, etc., according to point of view or direction of approach, cuts across both realms of living things. The branches of biophysics, biochemistry, paleontology, biogeography, etc., form bridges to the related sciences, so that a sharp delimitation of biology from other sciences from the point of view of method is not possible, either in the practical work of research or in scientific organization. Efforts to achieve a synthesis, particularly on the border lines of biology, are very active and successful at the present day.

II. Ways of Work in Biology

A. General Foundations

Corresponding to the character of biology as an empirical science, its statements have the general features of empirical statements. In the face of the great variety with which we are confronted by the world of living things and by the problems which this variety presents, the part played by pure description among biological statements is very great. The elementary constituent of a descriptive statement is the report of an act of observation—for example, the report of a color, of the measurement of a magnitude, of counting, of weighing, etc. It must, however, be remembered that even in these elementary descriptive statements the beginning of hypothesis construction must be recognized. The description chiefly serves for the characterization of a state of affairs in a way which will facilitate the recognition of other states of affairs of like kind. The observer will therefore emphasize, in the description of a fact, those sense impressions which seem to him essential and will neglect those which seem inessential. The manner in which a descriptive statement comes about is thus in a certain degree dependent upon the scientific attitude of the observer. This feature shows itself in an enhanced degree when descriptive reports are generalized and a fact is described not as unique but as "typical." Then the statement which is regarded as purely descriptive also takes on the character of a hypothetical statement. It is clear that this holds for all descriptions of processes which are characterized as following a typical course after repeated observations. Statements which express a functional connection between particular concepts—e.g., the statement "Cell respiration stands in a definite, regular connection with temperature"—in themselves exhibit the complete type of a hypothesis belonging to the empirical sciences. The concepts employed in them are defined by means of rules of operation, in which it is unambiguously stated how the practical observations are to be correlated with the concepts. In this way the concepts become constituents of the language of science, and their verbal meaning deviates more or less from that of everyday language. For that reason the words are often replaced by symbols which are unambiguously defined by the rules of operation. There is no sharp logical boundary between descriptive and functional statements. In many statements of biology descriptive and functional elements are mixed.

The general logical foundations of the formation of biological concepts and theories need not be discussed here in detail, since they hold for all the natural sciences and are therefore expounded in detail in other parts of this *Encyclopedia* (Lenzen, 1938; Woodger, 1939; Hempel, 1952).[2] For the same reason the characters which an empirical statement must possess if it is to be given a legitimate place in an empirical science will be only briefly enumerated. Such a statement connects a definite number of concepts in a manner which is free from objection from the formal logical standpoint. These concepts must be defined in a manner which makes their correlation with observable elements of experience possible. If this is done, the statement itself is testable by experience. By the making of an observation or by an experiment the statement can be verified or falsified. By 'verification,' here and in what follows, only confirmation by experience is to be understood, and by 'falsification' the absence of such confirmation. Nevertheless, by the use of these terms the question of the "truth" or truth-content of the empirical sciences is not touched upon. This question has no significance in biology other than that which it has in all other empirical science, and it is treated in other parts of this *Encyclopedia*. The uses of a statement of empirical science can also be stated in the following way: On the basis of such a statement, predictions about what is observable can be made. The occurrence of the predicted experience then means the verification of the statement, its absence the falsification of it. A correctly formulated hypothesis therefore has heuristic value in the development of science. On the basis of such a hypothesis new observations can be made, new experiments set up, which lead to the confirmation or falsification of the hypothesis. Hypothetical formulations with which observable processes, in the course of time, can be correlated and according to which such processes can therefore be predicted are called "rules" or "laws," e.g., Mendel's rules of inheritance. This terminology is also often chosen for formulations in which a relation between simultaneously existing observable elements is given, e.g., the law of the constancy of the chromosome number in species. The use of the word 'law' or 'rule' has often led to misunderstandings with which we shall be occupied later. This applies especially to the expression 'nat-

2. See Selected Bibliography at end for full data.

ural law,' now rarely used in biology. In principle we are free to construct hypotheses as we like—i.e., every formation and combination of concepts which satisfies the requirements sketched above is admissible. Psychologically speaking, the imagination of the investigator plays a great part in the conceiving of hypotheses and is frequently called "intuition." In practice, the conceiving of hypotheses chiefly takes place against a background of the state of the science already reached, having regard to the complex of statements which have already become more or less fixed constituents of this science.

Statements about biological facts which according to their structure could be testable by experience but which for practical reasons cannot at the time, or perhaps ever, be tested are admissible as hypotheses even though no judgment about their confirmation is possible. Statements which according to their structure are not testable by experience and which, on account of the absence of an unambiguous correlation between the symbols and the observable facts, can never in principle be verified or falsified have no function in the system of an empirical science. They are neither false nor correct in the sense of empirical science but cannot be used in science and have no legitimate place there. Whether they are, in general, scientifically meaningless is a question which will not be discussed here. But there is still another type of statement whose confirmation is in principle untestable. These are statements which, although externally they have the form of a hypothesis of the empirical sciences, are in fact of such a nature that the rules of correlation between the concepts used in the statement and what is observable constitute the whole content of the statement itself. These statements are occupied exclusively with repeating the rules of operation. These are tautologous statements. If in the statement "The positive phototactic reaction of a *Euglena* is proportional to its light-requirement" the concept "light-requirement" is only testably defined by means of the establishment of the behavior under the stimulus of light, this is a tautologous statement of the above kind. Such tautologies have no direct value for science. Nevertheless, as will be discussed later, tautologous statements frequently occur in scientific expositions.

If a hypothesis is falsified by a reliable observation, it is abandoned. If it is confirmed by the observation, it is used further in science. A hypothesis which is frequently verified comes to be regarded more and more as a stable constituent of the science and often becomes a habit of thought. If a hypothesis can be tested in various ways in accordance with the rules of correlation and if all these tests verify it, then the hypothesis is regarded as an especially well-confirmed one. If the incorporation of a new hypothesis in the permanent structure of the science takes place without contradiction, this is regarded as a proof of the "correctness" of the hypothesis or of the "truth" of its content. If the incorporation does not succeed without contradiction, then, by a thorough logical analysis of the contradictory statements and their elements, we must investigate whether the contradiction is not merely apparent and whether it cannot be removed by a logical change. In other cases the contradiction can be bridged over by means of accessory hypotheses which restrict or extend the validity of the hypothesis and which, in turn, must satisfy the requirements of testability by experience. Naturally, they must not be introduced only *ad hoc*, i.e., they must not be merely tautological or formulated without any connection with the rest of the system of the relevant science, because in that case they would in principle be removed from any testability. On account of their heuristic value they can often give birth to a new development in science. If a far-reaching contradiction persists between old and new hypotheses, this leads to a "crisis" in the empirical science concerned. This, in turn, leads to a revision of the system of statements hitherto in use. It may then prove to be the case that certain theories that have hitherto been accepted must be abandoned because they no longer correspond to the augmented experience or that the old theory represents a special case of the more comprehensive, newer theory. Growing experience leads to an enrichment of our stock of scientific possessions and necessitates a perpetual revision of the complex of statements of the science, and in this lies its progress. The contradiction is never in the experience but always only in our formulations. Empirical science knows no "aporia" in the philosophical sense.

If, by the observation of a phenomenon, an existing hypothesis is verified, we say that the phenomenon is "explained" by the hypothesis, especially when we have to do with a frequently confirmed hypothesis. The "explanatory value" of a hypothesis seems to us to be especially great if many different experiences can be correlated with it and if it can be added to the whole edifice of science without contradiction. Some statements are formulated or interpreted with the intention of making a process or a complex of relations "intuitable" or "picturable" to human beings. Many formulations of natural science, to which a high explanatory value is ascribed, are so abstract that this feature of intuitability is entirely absent. The difficulties and misunderstandings which can result in science from the human need for intuitability and from the confusion of "explanation," in the sense here defined, with "intuitability" will be discussed later with examples.

In the following pages an attempt will be made to show that the statements of biology in its various subdivisions satisfy the requirements described above and that biology can be regarded as a system of such statements. According to the way of considering and of investigating biological data, we shall here carry out a twofold division into elementary and complex points of view. This subdivision is purely practical and is intended to make our methodological exposition easier.

B. *The Elementary Points of View*

In considering an organism, we can direct our attention, according to our inclination and aim, either to its visible peculiarities or to its behavior and performances or to its significance as a member of a reproductive chain, and in accordance with these possibilities it is customary to distinguish the subdivisions of morphology, physiology, and genetics. The individual investigator will, according to his gifts, his training, or his particular problem, direct himself exclusively or predominantly to the one or the other point of view. But, in spite of the necessarily high specialization of the methods involved, it is naturally impossible for him to devote himself exclusively to the one or the other point of view, so that here also the boundaries are not sharp. In dealing with complex problems, the synthesis of the various points of view is even unconditionally necessary.

1. The Morphological Point of View

The enormous multiplicity exhibited by living things necessitates at the outset a description of the visible "characters" of particular organisms—their forms, proportions, colors, and measurable magnitudes. This morphological description relates not only to the external, but also, in the form of anatomy, to the internal, structures, the construction and mutual spatial relations of the organs. It continues right down to a morphology of the tissues, of the cells and their parts, in the branches of histology and descriptive cytology. An essential feature of descriptive morphology, in the widest sense of the word, is the discovery that with its help a classification of organisms is possible, that their recognition on the basis of a sufficient description is possible. The process of classifying in this way involves at the same time the setting-up of a hypothesis—namely, the conception of the "type" of the organism concerned, by means of which we can undertake, in the particular case, to correlate the objects found in nature with the appropriate type in a manner sufficient for practical purposes.

To begin with, the expression 'type' is to be understood in the widest sense of the word. The process of setting up types occurs in every morphological as well as physiological classification, for the reason that the descriptive elements used for the purpose never merely represent the description of an observed result but are constructed from a series of such results by abstraction of the features which are common and which seem essential to the investigator. Where the process of classification into types is carried out in a deliberately critical spirit by the investigator— say, in connection with genetical or biometrical problems—certain conventional rules are set up for carrying it out. For example, a classificatory measure of size is given as the mean value of a definite number of measurements, with its mean error and the width of variation of the character given by the standard deviation. A pattern of markings is arranged according to a series of standardized categories, and so on. But such a strict interpretation of the method of classifying into types is usually avoided, and we content ourselves with specifying such characters as will satisfy practical requirements. The most diverse categories which are descriptively specifiable can be the object of the typifying process. It may be a race, a species, a higher classification unit, a group of organisms brought together from some standpoint other than the classificatory (e.g., all water plants), which is described and so made an object of the typifying process. The species concept of taxonomy is thus only a special case for the delimitation of which, as will be discussed later, other than classificatory notions are required. But an organ, a tissue, a cell form, a metabolic process, a mode of instinctive behavior, etc., can all be objects of the typifying process and in this way undergo a classification. Not only do we work in this way in the morphological branches of biology, but the description of physiological processes on a comparative basis leads to a classification by a like process.

The establishment of the "type" of a form, of a mode of behavior, or of a process involves the formation of a hypothesis, in so far as an arbitrary selection of results of observation is compared and emphasized as essential for classification, and thereby the hypothesis of a regular, common repetition of these characters suitable for classification is set up. In this way a law of coexistence of characters, which is empirically testable, is asserted. To the comparison of results of observation there is, of course, added, in this procedure, an idealization, brought about by the fact that the choice of what is described is to some extent arbitrary. This idealization appears in increased measure in the working method of comparative morphology when this proceeds to the setting-up of higher types of higher order. By the comparison of types of similarly organized living beings, we reach in comparative morphology the highly abstract concept of the "constructional plan." In this comparison similarities and differences are established, and essential characters are distinguished from inessential ones; and in this way a "fundamental" constructional plan, common to several organic types or a large group of organisms, is constructed. By further comparisons more and more comprehensive, more general plans of construction can be set up. The bodily parts, organs, and parts of organs which correspond to one another in the constructional plan are called "homologous" structures.

By comparing the structural plan, say, of a vertebrate, with the concrete natural objects, e.g., with a horse, a whale, and a bird, the procedure of "homologizing" is carried out. The anterior pair of legs of the horse is homologous with the anterior fins of the whale and with the wings of the bird. Homologous organs are thus such as correspond to one another in their place in the structural plan, those which can take the place of one another in the structural plan and which coincide in the plan of higher order. Homologous organs exhibit a construction out of fundamental elements which correspond to one another—e.g., homologous parts of the skeleton, of the musculature, of the innervation.

The formulation of the concept of homology has the typical form of a hypothesis of empirical science. It represents a system of schemata, the structural plan with the spatial relations of characters and organs which are characteristic of it, and it can be empirically tested in so far as the presence of the homologous organs can be verified in every concrete organism by a morphological investigation. It permits predictions of the kind which assert that, if an organism shows itself to be a vertebrate by one characteristic feature, then we may also expect the other homologous structures of this group. Since the domain of validity of the hypothesis is restricted to a particular group of organisms, we should not, in the case of a falsification, place the organism under investigation in this group, i.e., the Vertebrata. It will "correspond" to the structural plan of another group.

In spite of the high degree of idealization which is involved in the setting-up of structural plans of high order and in an extended use of the concept of homology, the statements of comparative anatomy are nevertheless empirical statements. The elements for the construction of structural plans are derived from experience, and the concepts used for this purpose remain, through their definitions, empirically testable. In this way the extreme abstractions of morphology, in so far as they remain scientifically useful, are distinguished from the entities of geometry, the construction of which takes place on the basis of a purely intellectual conception and for which the requirement of testability by experience is meaningless.

As an example of a morphological statement of wide scope, we may mention the law of the constancy of the chromosome number for any species. It holds for all organisms (with the exception of microörganisms which have no nucleus) and states that in all mitoses of a member of a group defined as a species (we shall return to this definition later) the same number of chromosomes is always to be observed. Instances of the falsification of this rule require the setting-up of testable accessory hypotheses, e.g., the hypotheses of chromosome races, of polyploid races, and the like.

Statements like the above are reached by a process which M. Hartmann (1948) calls "generalizing induction." The verification of a statement by an experiment he calls the method of "exact induction." These expressions do not seem to me to be well chosen, since an essential methodological distinction between the two procedures cannot be maintained on closer examination. It would, perhaps, be better to contrast the two procedures as "comparative" and "experimental." It is by no means the case that obtaining and testing statements by comparison is especially characteristic of the morphological point of view, let alone the only one open to it. A large part of morphology is called "experimental" morphology because the method of obtaining and testing statements is that of experiment, and thus "exact induction." The statement "A certain plant forms finely divided leaves under water but on the surface has undivided floating leaves" can be tested not only by observations on cases occurring in nature but also by experiment, as by artificially setting up the conditions mentioned and seeing whether the predicted connection is realized. This statement connects concepts descriptive of the environment with morphological concepts.

There is a tendency to assign to experimentally testable statements a greater "demonstrative power," or at least a greater picturability, than is assigned to statements which are testable by comparison. It is customary to contrast them, as "causal-analytical" statements, with "systematic" or "order-analytical" statements. The expressions 'causal-analytical' and 'order-analytical' are not very fortunately chosen. In the first place, it is a question not of analytical statements in the logical sense but of synthetic ones. Moreover, neither a causal connection nor an ordering is "analyzed." It would be more correct to say that these statements set up rules or laws according to which in the first case a succession of states in time, in the other case a co-existence of characteristics or a correlation between properties which is empirically testable, is predicted. We could—with certain reservations—also speak of "the search for a causality of process" or of "a search for an order." If we wish to restrict the concept of "order" to the establishment of a law of coexistence of properties, then only those statements are to be called "order-analytical" which are formed on the basis of repeated observation of such a coexistence and are formulated so as to be testable on further observations of the same kind: for example, statements about anatomical structures or about the results of bio-chemical analyses. If a statement is formulated on the basis of repeated observations of processes in which one particular order passes regularly into another, then this statement is also obtained by comparison and is testable by further comparative observations. Nevertheless, in such cases it is customary to speak of "causality of process," and we should have to regard these statements as causal-analytical ones, although they have been obtained by comparison. Experiment is then a special case of this procedure in which we deliberately bring about a particular initial situation in order to test the succession of changes of state demanded by the hypothesis. By arbitrarily varying the initial situation, experiment opens up new possibilities for the setting-up and testing of hypotheses and in that way leads to the enrichment of science. Since the expressions 'causal-analytical' and 'order-analytical' are much used in the literature, they will be retained in the following pages, in spite of the reservations just mentioned.

Both types of statements are genuine statements of empirical science if they satisfy the requirements sketched in Chapter I. The changes of state in dead, and especially in living, nature to which our experiences relate bring it about that we are often placed under the necessity of transforming order-analytical statements into causal-analytical ones. In this way one established order is traced back to another, and a rule is set up for the transformation of one arrangement into another which is empirically testable. It is a question of great importance from the point of view of the theory of knowledge whether, in principle—if only under certain, perhaps not always given, assumptions—*every* order-analytical statement can be transformed into a causal-analytical formulation. The question then arises about the "final order" to which everything can be referred back. Since this and similar fundamental epistemological questions are common to all empirical sciences, and biology occupies no special position in relation to them, they will not be discussed here. In any case, it very often happens in biology that order-analytical statements are transformed into causal-analytical ones and that attempting to do this is of great heuristic value.

In the field of morphology this is the case when questions of form and structural plan are expressed as questions of formation and development, in the ontogenetic or phylogenetic sense. M. Hartmann (1948) has expressed the opinion that the "exact inductive" method is superior to the "generalizing inductive" method and, by way of example, compares the following statements: (1) "All carnivores have a relatively short, all herbivores a relatively long, gut" (generalizing inductive); (2) "On an animal diet tadpoles acquire a relatively short gut; on a vegetable diet they acquire a relatively long one" (exact inductive). The second statement, which can at any time be verified by experiment, carries more conviction, according to Hartmann, than the first. W. Zimmermann (1948) on the other hand, objects that the example is not well chosen, because, for testing the efficiency of the two methods, two very nonequivalent facts are compared. In the case of the tadpole we are dealing with an ontogenetic developmental process which is dependent upon diet; the example from comparative anatomy has a scientific meaning only if it is formulated as a phylogenetic question. The assumption must then be made that, in adaptation to dietetic conditions, among the various animal species the different relative lengths of gut have in the course of phylogeny become species-specific, genetically fixed characters. I agree with Zimmermann that in the examples chosen by Hartmann we have to do with very different facts, but I am not of his opinion that the first statement, which belongs to comparative anatomy, must be denied all scientific value, inasmuch as it is not formulated as a phylogenetic statement. As a pure order-analytical statement, it satisfies the general requirements of an empirical statement and at the same time calls attention to an existing regularity. If we wish to transform it into a phylogenetic statement, then we assume that similar statements can be formulated about all processes occurring in phylogeny and that these would be—at least in principle—empirically testable. But then the statement is transformed from an order-analytical one into a series of causal-analytical statements.

A greater measure of picturability attaches to the concepts and formulations of morphology than is the case for the statements of other branches of biology. In using these concepts, people are often tempted, in consequence of this picturability, to choose a linguistic expression which, although very picturable, nevertheless does not, strictly speaking, agree at all with the logical foundations of empirical science. This is so, for example, when it is said that in the development of an organism a structural plan is "realized," that the development "deviates from the structural plan," or that it "follows the structural plan" or "is governed by it." The dubiousness of this mode of expression becomes still clearer when we speak in comparative morphology of the "metamorphosis of homologous organs," of a "transformation of shoots into thorns," or of a "transformation of the various biting mouth parts of one insect into the corresponding differently formed sucking mouth parts of another." Everyone who is scientifically educated naturally knows that such linguistic expressions are intended only in a metaphorical sense. The "structural plan" exists, indeed, only as an idea in our consciousness, and the "metamorphosis" of homologous organs is also only an imaginary process. The case is somewhat different when we speak of "ontogenetic" or "phylogenetic" developmental processes, but then other rules hold concerning the correlation with the facts of experience. In the picturability of the notions often used in the discussion of morphological problems there lies a danger of falling into a realism of ideas which, as a legacy of the traditional philosophy of nature in the form of idealistic morphology, has played a great part in the historical development of biology. Some modern biologists do not seem to be quite consistent in avoiding this danger, and some profess quite openly a more or less concealed realism of ideas (see Chap. III, B, 2–3).

There is no sharp delimitation between the morphological and the other points of view in biology. The simplest descriptive data often point to the function of the organ or other physiological relations. Every anatomy, especially every histology and cytology, is in some measure functional anatomy and so leads into physiology or forms its foundation. The morphological point of view has special relations to the problems belonging to systematics and evolution which we shall discuss among the complex points of view.

2. The Physiological Point of View

Physiology is commonly defined as the theory of the functions of the organism, of its organs and tissues, or, better, as the theory of the processes which take place in the organism and between the organism and the environment. Sometimes we are told that only physiology deals with the material and energetic properties of organisms, morphology dealing with their forms.

A statement about physical forms, however, has a meaning for empirical science only if it is testable by experience and hence on material structures. Everything independent of "material" carries with it the danger of a realism of concepts. If a form or structure concept in a biological statement is defined in a way which is completely independent of the "substance," the whole statement loses its testability by experience and with that its usefulness to empirical science. Other assertions of idealistic morphologists teach that causal-analytical thinking is characteristic of physiology, while for morphology we have "prototype" thinking. Although the notion of "prototype," conceived as a kind of intuitive picture, is not defined so as to be testable in experience, yet idealistic morphologists try to set up a kind of causal relation between the prototype and the empirically testable forms and structures of organisms. We shall return later to these differences of view regarding the concept of cause (Chap. III, B, 3).

It is also incorrect to say that physiology is the domain of "exact induction," while morphology is the field of "generalizing induction." Moreover, many statements of physiology are constructed of purely descriptive elements and are obtained by comparison and thus order-analytically. Again, physiology provides many examples of the possibility of transforming order-analytical statements into causal-analytical ones. Physiology often works with purely physical or chemical methods, as in demonstrating the transformations of materials in metabolism or in the analysis of the processes in semipermeable membranes. It frequently expresses its statements in formulas of a high degree of abstraction in which quantitatively expressible functional relations between two or more variable magnitudes are asserted to hold. On account of this structure such formulations are regarded as especially "exact," as strictly scientific or demonstrative. They are, however, not more and not less demonstrative than every other empirical statement which conforms to the requirements of the logic of science.

The following is chosen as an example of the properties of physiological statements described above: "The respiratory quotient CO_2/O_2 is approximately 1 in herbivores, somewhat less in omnivores, and still less in carnivores." The descriptive elements of the kind of diet of various organisms and their gas exchange, used in the statement, can be defined by specifying the particular diet and by giving methods for the exact determination of the gas exchange. In this way unambiguous correlation of the statement with the testing facts is made possible. The stipulation about the gas exchange can be formulated quantitatively, as well as the physicochemical composition of the food, if necessary. The whole statement has the character of one obtained by the generalizing inductive method and thus of an order-analytical statement. But, from it, single statements can easily be derived which have the character of experimentally testable causal-analytical statements. This becomes especially clear if we set up more restricted statements about the utilization of nutrient materials in metabolism—for example, about the utilization of carbohydrates, fat, and protein as respiratory material—and in this way connect a whole series of testable causal-analytical statements, which make the rule expressed in the wider statement easily understood. In this way the statement shows itself to be a "blanket statement"[3] about an exceedingly complex occurrence which we can—in this case—analyze quantitatively, and to a great extent, into its partial processes. In consequence of this we regard the process as especially well "explained," since we can, from the more comprehensive statement, derive a diversity of statements which are testable by various methods and which can be connected with one another and with other experiences without contradiction. In this way single statements of a particular kind, e.g., those about the utilization of proteins in metabolism, can be related to more comprehensive statements, e.g., about the respiratory processes.

Let us take another example from the physiology of irritability. The phototactic behavior of *Pandorina*, a free-swimming

3. [The term 'blanket statement' is a translation of *Pauschalaussage*. In correspondence the author explains this term as follows: "By 'Pauschalaussage' I understand a simple statement, the background of which is a complex occurrence, and which can therefore be expressed—under certain circumstances—as a complex of other statements."—EDITORS.]

colony-forming green alga, is characterized by a lower threshold, i.e., the light-intensity at which the first positive phototaxis is observed and below which no reaction occurs; the turning point, i.e., a stronger light-intensity, at which the positive phototaxis turns into the negative; and the upper threshold, i.e., a still stronger light-intensity, at which the negative reaction is also absent. The position of these cardinal points of phototactic behavior is dependent in a definite way upon the hydrogen-ion concentration of the solution in which the colonies live. Here the concepts which are connected in the statement are, on one side, a physical property of the environment (the light of a particular intensity coming from one direction), on the other, a mode of biological behavior (the unilaterally directed movement of the alga colony, which is produced by a particular sequence of strokes of its flagella). The whole statement is thus undoubtedly a blanket statement about a very complex occurrence within the organism, extending from perception of the stimulus up to the response. This blanket statement can also be analyzed into a series of testable single statements, e.g., a statement about the place of perception of the stimulus and its structure, one about the conduction of the stimulus, one about the structure of the flagella and their function, one about the changes of state of the whole reacting system which are connected with the changes of the hydrogen-ion concentration, and so forth. The light is then called the "stimulus" (or, better, "cause of the stimulus") and the behavior of the organism its "answer" (or response) to the stimulus. The dependence of the position of the cardinal points on the hydrogen-ion concentration is described as a *Stimmungsphänomen* ("phenomenon of mood") and the shifting of the cardinal points according to experimental changes of the hydrogen-ion concentration as *Umstimmung* ("change of mood") of the organism. This simple example from the physiology of irritability illustrates the point that, in formulating statements about such processes, linguistic expressions are frequently used which are taken from the psychology of human experience. In this way statements become more "picturable," without a special "explanatory value" being given to these statements by the use of such linguistic expressions. The possibility of empirically testing them remains instead just the same as for every other physiological, i.e., empirical scientific, statement. If we wish to define the concept of "mood" (*Stimmung*) in a scientifically useful way, we can do this only by correlation with definite experimental observation, and the same holds for the concepts "stimulus," "stimulus-perception," etc.

The use of anthropomorphic terms from human psychology is widespread, especially in the whole of the physiology of the nervous system and special senses of the higher animal organisms, including man. With a consistent application of fundamental scientific criticism, no difficulties result from the use of these linguistic expressions. Even the complex behavior of higher animals can be treated throughout as a problem in purely empirical science. But that a great danger lies in these anthropomorphic modes of expression, which can lead to a conscious or unconscious psychologizing of biological facts, will be shown later (Chap. III, B, 2).

Only in the case of man does a really new problem arise—from the fact that to him is given a new kind of experience, the inner experience, in the form of his own feelings and sensations. One and the same state of affairs can be described in the form of neurological statements and thus in accordance with empirical science and the requirement of experimental testability, or it can be described as sensation in the expressions of psychology. In this way arises the complex of problems of the correlation of physiological with psychological experiences, the psychophysical problem. Since this cannot be discussed without a fundamental critical analysis of psychology, we shall not deal further with it here. It need only be pointed out that an excellent and unambiguous clarification of the boundaries between the physiological and the psychological points of view from the biological side has been carried out repeatedly in recent times (e.g., Bünning, 1949).

A special role is played by the physiological standpoint in the complex problems of developmental physiology, in ecology, and in the investigation of behavior.

3. The Genetical Point of View

In opposition to the usual scheme, the fundamental point of view of genetics will here be given a separate treatment, since it shows, in practice, a certain independence in biology and plays a part in all complex problems of biology, on account of the great importance of modern genetics. The geneticist does not consider, as does the morphologist, the structure of an organism or those of several distinct organisms comparatively; neither does he, in the first instance, investigate the processes which occur in organisms, as does the physiologist. The geneticist deals comparatively with the individuals of a population which form a continuing reproductive chain.

The word 'heredity' (*Vererbung, l'hérédité*), which is used in genetics in order to denote the reappearance of characters of the ancestors among the descendants, conceals within itself the danger of misunderstandings which even today still often lead to senseless debates. It is taken from the language of everyday life, in which it signifies a process in which an external possession passes from ancestors to descendants. Here we have actual possessors, e.g., father and son, and the inherited object. In a derivative sense we speak in genetics of "inheritance" and of "inheritable properties." The word 'property' is used here as though there were a "bearer," independent of the properties, by which these properties are inherited. Nevertheless, no scientific statement can be formulated in which the notion of the bearer is so defined that it could be tested in experience according to definite rules of correlation, independently of all properties. On the contrary, it is always the properties themselves, and these only, which, in accordance with definitions, can be empirically tested. The organism itself is definable only by means of the properties which serve to characterize it. The idea of a bearer independent of the properties is an analogy with the idea of self among human beings, and its transference to things and organisms is an anthropomorphism, or conceptual realism. In the scientific language of biology, therefore, 'inheritance' and 'inheritable property' mean only that, in the comparison between ancestors and descendants, like or deviating characters can be established for the classification of the individuals. The comparative method leads to the setting-up of statements of the order-analytical type. To see in the connections so expressed a process of inheritance implies a further hypothesis, through which the order-analytical statements are transformed into causal-analytical ones.

The best-known general statements of genetics are the Mendelian rules or laws of inheritance. In the statement "After crossing a bean with red flowers with one with white flowers, we find (after self-fertilization in the F_1 generation) in the F_2 generation red-flowered and white-flowered plants approximately in the proportions $3:1$," there is expressed, in the first place, on the basis of the comparison of the results of repeated experiments, a definite correlation between the characters of the parents and those of the succeeding generations. Owing to the fact that the organisms compared belong to a reproductive chain, the causal-analytical treatment of this relation is appropriate. This was formulated by Mendel in the form of the hypothesis of the hereditary factors. In its later form this is composed of the following statements: "The factor R 'produces' the red, the factor r the white, flower color"; "Every plant possesses a pair of hereditary factors, either RR or rr or [the F_1 hybrid in the above example] Rr"; "Whenever the sex cells are formed, the members of these pairs are separated, and each sex cell possesses only an R or an r"; "Through random combination of the gametes in fertilization, the ratios of the red- and white-flowered plants in the F_2 generation are explained." The assumption that a factor R "produces" the red flower color is conceived after the pattern of the "forces" of physics and, to begin with, means no more than that, in this way, a correlation is proposed between the hypothetical concept "factor" and something observable. If the statement were restricted to this correlation, i.e., had no other content than this rule of operation, then it would be a tautology and therefore not testable by experience. But, owing to the fact that the concept "factor" is invested with still other defining characters which are empirically testable, such as its behavior during gamete formation and fertilization, the whole hypothesis becomes scientifically useful. It asserts something about hypothetical units which are distributed and about a hypothetical mechanism which distributes them. The heuristic value which the hypothesis possesses is shown by the various experimental arrangements by which it can be tested, e.g., the backcrossing of the F_1 hybrids with the dominant or the recessive parent, the analysis of the F_2 generation by crossing with the recessive grandparent, the testing of the assumption of the purity of the gametes in the hereditary processes in the case of haploid parthenogenesis and haploid organisms, the testing of the mechanism of distribution by tetrad analysis, and so on. Beyond this, the heuristic value of the Mendelian hypothesis has proved itself precisely in those cases which seemed to falsify the hypothesis. The setting-up of testable accessory hypotheses has empirically disclosed new connections and, just through these cases, has led to a significant extension of knowledge. We feel that observations about inheritance are "explained" (for instance, the relative frequencies in the F_2 generation) when they can be derived from the general hypothesis. By combining the methods of hybrid analysis with the cytological and embryological methods of modern genetics, such observations can also be derived from yet other hypotheses, such as those about the minute structure of the chromosomes. Through the consistent combining of all these observations, the explanatory value of the hypotheses involved seems to us to be especially satisfactory.

The Mendelian rules are statements about the relative frequency of the various hereditary types in the generations following a hybridization; they are thus statements about the structure of a collective, i.e., statistical statements. The Mendelian hypothesis of the hereditary factors thus characterizes the distributive processes, assumed in gamete formation and in fertilization, only in a statistical way. Naturally, this does not mean that these processes in the individual case, and consequently the individual hereditary fate of a member of the F_2 generation, are indeterminate—say, in the sense of microphysical uncertainty. All these processes which decide the fate of individual descendants after hybridization are, in fact, definite macrophysical processes and could be described in a determinate way by other formulations for the individual case, although this would be very difficult in practice. Statistical statements also occur in other ways in genetics, as in the domain of the statistics of variation. All statistical statements require for their empirical testing a treatment of the observational material by the theory of errors. This is a general requirement of scientific method, has nothing to do with the object of the investigation, and therefore does not mean, as some biologists believe, an encroachment of methods not belonging to biology. The frequent use of statistical statements has, especially among some theoretical authors, given rise to remarkable and erroneous judgments about genetics. One observer sees, for example, in the Mendelian rule, because it appears in mathematical form, the ideal example of a biological "law," while another, for the same reason, would deny it every biological character. The lack of reflection from the point of view of the logic of science is especially noticeable in such discussions.

The concept "hereditary factor," which governed the early days of genetics after the rediscovery of the Mendelian rules, has, in the course of the development of this science, merged into the wider concept of the "gene" or "allele." The notion of the gene, as it is used today, is a good example of biological theory and concept formation. Today the gene is defined by so many points of determination that a great number of different testable formulations can be derived from statements in which this concept occurs. In the experiment of hybrid analysis the gene appears as a segregating unit, in cytogenetic experiments as a localization unit, in mutation experiments as a unit of mutation, in embryological experiments as a unit of action. Nevertheless, the concept of the gene is by no means dogmatically frozen, as outsiders sometimes believe, but is in a state of constant change, and a series of heuristically valuable tendencies in modern genetics points the way to a revaluation, perhaps even to a fundamental revision, of this concept.

During recent years the genetical point of view has occupied an increasingly important place in biological research and continues to gain ground. It seems to be destined to form a connecting link between the other aspects of biology and at the same time to promote a synthesis of biological thinking. It therefore plays a large part in all complex questions of biology, especially in the domain of developmental physiology and, in the form of the genetics of populations, in the questions of systematics and evolution.

C. The Complex Points of View

In this section we shall try to show by means of some examples how statements about complex states of affairs are formulated in biology. Every simple biological experience can be represented in various ways on the basis of the elementary points of view described in the previous section. In the case of more complex situations a synthesis of the various points of view is unavoidably necessary. The subdivision of this section in accordance with some fundamental problems of biology is quite arbitrary and in no way exhaustive.

1. The Organism as an Open System

The attempt to reach more general statements which will be true of organisms on the basis of the various points of view leads, among other things, to the definition of the organism as a system. A system is a structure in which all processes are connected functionally in a more or less complicated way. The rules which assert something about *single* processes occurring in a system then only hold conditionally, with reservations, since in such cases the mutual relations of the processes considered to all other processes are neglected or deliberately simplified. Almost all biological statements struggle with this difficulty. An exhaustive total representation of the mutual relations prevailing in a complex system by means of a single statement is not possible. The more general and comprehensive such a statement is, the more indefinite are the concepts used in it and the less testable are they in experience. That every organism is a system is a very general statement, which must first be supplemented, in every single case, by a whole series of special statements if it is to be testable in experience.

There are systems in nonliving nature also, so that the system character of the organism does not provide the means for a strict delimitation of the living from the nonliving. Naturally, the relations in living systems are for the most part of a much higher degree of complication. Moreover, the nonliving systems are not, in principle, closed systems, although most of them can, in practice, be regarded as such. This is not possible with organisms. On account of the persistent mutual exchange of material and energy with the environment, organisms are open systems to a great extent. All possible changes in the environment can have far-reaching effects in the system. The attempt to treat the living organism experimentally as a closed system leads to its destruction, to death. In consequence of the open-system character it is still more difficult in the case of organisms than it is in nonliving systems to draw a boundary between the system and the environment. Every biological statement, strictly speaking, not only relates to the organism but takes in a part of the environment as well. The "thinghood" of a living being is only a psychological experience of the human observer; in a strictly scientific statement the organism, or one of its parts, occurs never as a completely closed definable "thing" but only as a conventionally marked-off part of what is accessible to empirical science. From the latter point of view it is therefore not meaningful to wish to draw a boundary between organism and environment or to distinguish in principle between living and nonliving parts or substances in the organism. The materials and energies which enter the organism from the environment take part more or less intensively in the processes occurring in it, and the same may be said of those passing out of it. It cannot be meaningfully stated when these chemically or physically definable things are "alive" or when they are not. Naturally, for practical purposes, we can often make a delimitation between organism and environment which is exact enough for the purpose in hand. Similarly, we can often distinguish between living and dead parts of an organism in a way which suffices in practice.

The processes occurring in organisms are so connected with one another that, in spite of the perpetual change of the materials and energies composing the system and in spite of the most diverse disturbances and variations of state in the environment, the actual individuality of the system in question is, within certain limits, preserved or is restored after being disturbed. Only if the disturbances by the environment or the processes in the interior of the organism exceed certain limits is the system irreversibly destroyed. This ability of organisms has been called "self-regulation." The state in which a living organism finds itself is a state of equilibrium (variable within limits which are characteristic for each organism) in which, in contrast to the forms of equilibrium prevailing in the practically closed inorganic systems, a whole series of processes occurs in disequilibrium—a state which has therefore been called "dynamic equilibrium" (Hopkins) or, very appropriately, "flow equilibrium" (Bertalanffy). Many complexes of biological statements of fundamental importance contain more special statements about equilibrium states or differences of potential of the most diverse kinds—for example, statements about the function of the enzyme system in living cells, about the osmotic relations in cells, about changes of state in the conduction of nervous impulses in nerves, and so forth. Such statements need not always owe their origin to the physiological point of view; statements about states of equilibrium also play a part in the morphological point of view.

The most general statements about the state of living systems are especially concerned with the energy relations in the living world. The maintenance of the flow equilibrium or, in other words, the preservation of the potential differences necessarily present in the living organism, is connected with incessant performance of work; the organism raises in this way the entropy level of its environment; it feeds, so to speak, on "negative entropy" (Schrödinger). In connection with such considerations and similar ones, it is often customary to speak of the "activity" of the organism as one of its special characteristics. So long as this word is used only to denote the mutual energy relations between organism and environment, it is meaningful from the point of view of empirical science. Unfortunately, the use of such words and ones similar to them often leads more or less consciously to anthropomorphic ideas, like the ideas of active "forces" in physics, which have long been superseded.

In order to make the complicated system relations in organisms more easily intelligible, recourse is often had to so-called "models." In biology experimental arrangements or theoretical constructions are called "models" if they show, with deliberate simplification, one or more far-reaching analogies with the processes in the organic world. As examples, we need only mention the experimental permeability model of Osterhout, the theoretical derivation of the average cell size and of cell division from the data of metabolism by Rashevsky, and finally the molecule representation of the gene. Such models are not hypotheses in the usual sense, since they cannot be directly tested by biological observation in the individual case. Rather, they are complexes of statements which have already been verified by physical or chemical experience and are applied to biological data almost allegorically with conscious restriction and certain reservations. They share with the biological hypotheses the property of heuristic value. The model serves as a general pattern in accordance with which special hypotheses about processes in organisms are constructed which are then directly testable by biological observation. A misunderstanding of this function of models in biology may easily lead to the superficial opinion that through them a series of important properties of organisms is "explained."

Among the various mutual relations both in the structure and in the function of living systems, it is possible to distinguish more or less clearly those of greater from those of lesser importance: a kind of hierarchical order prevails within these mutual relations, which is also called "organization." This feature is also not entirely absent from nonliving systems, yet in living systems it shows a far higher degree of complication. In this general form the statement about the "order" prevailing in living systems has only a very vague scientific meaning. If we give it a purely physical meaning, then it is a statement about the uncommonly high degree of atomistic order which we find in the structure of an organism in contrast to its environment and which, in fact, occurs nowhere else in nature. If we wish to make use of the statement about the given order for the clarification of biological facts, then we must set up more special statements about the kind of this order, its rules, and their correlation with experience, in order to make it testable by observation. Failure to attend to this requirement often leads to the purely tautological use of general statements about order in the organic world.

It is not only to the whole organism that the notion of the open system can be applied; this notion is also applicable to the parts. This holds especially for the cell, which, even in the union of the many-celled organism, in both its structure and its functions clearly shows the properties of a complex open system and is therefore rightly called an "elementary organism." Since more has become known about the cell nucleus, especially about the highly complicated structure and function of the chromosome apparatus, these structures must also be given the character of complicated systems. The morphological and physiological points of view lead to a complex of statements about the system relations prevailing in the organism among its organs, cells and cell parts, and the processes taking place in them. The standpoint of genetics reveals in the genome a system of special order, which is a guiding center for all these processes and is therefore directive for the establishment of the properties and modes of reaction which can be demonstrated morphologically and physiologically in the organism. The rules for the causal connections between the system of the genome and that of the organism form further complexes of statements of a genetico-embryological point of view. With the high degree of complication which statements acquire in this way, it is understandable that every hasty generalization or simplification jeopardizes the scientific value of such statements and that only the strict requirement of empirical testability guarantees the wider utility of such formulations. For the most part, this requirement is satisfied in the domain of modern genetics in an exemplary manner.

The organism is connected with the environment by a series

of complex mutual actions. If we direct our attention to this, as we do, for example, in ecology or in the study of the behavior of the higher animal organisms, then we consider the organism and the environment, or a particular portion of the environment, as a system. In the biocoenoses which inhabit a particular environment, various organisms stand in regular reciprocal action with one another. In an increased degree this holds for closer relations between different organisms, as these are given in the various forms of parasitism and symbiosis. We can in this way extend the system notion to systems of higher order in which equilibrium states with certain capabilities of self-regulation can likewise be established. What has been concluded above from a consideration of the single organism as a system holds also for statements belonging to these domains of biology in all essential respects.

The high degree of complexity which belongs to the systems of higher or lower order considered in biology brings it about that many of the statements of biology are blanket statements. These establish a regular relation, for example, between the nutrient materials taken in and the end-products of metabolism given out or between the temperature and the formation of a pigment in the body covering or between the moisture of habitat and the occurrence of a particular plant species or between the development of a parasite and the conditions in its intermediate host, and so on. Such statements deliberately refrain from describing in detail all those processes which must be assumed as part-processes in the domain of facts under consideration. They are, nevertheless, typical statements of empirical science. The great frequency of such statements of quite a special kind distinguishes biology from other branches of natural science. Nevertheless, it is not correct to regard only statements of this kind as genuine "biological" statements and to deny this title to statements which deal with single processes in the organism and therefore are or have the type of physical or chemical statements. There are in biology no "more or less biological" statements but only statements belonging to empirical science.

2. Growth, Development, Reproduction: The Historical Character of the Organism

It is a peculiarity of living systems that many processes in them, including very essential ones, are irreversible, or run in only one direction. Every organism shows a chain of changes of state following one another in a manner which is characteristic for the organism. From a plant seed, if it does not die, a plant will grow which, if it does not die, will form seeds in its turn. An insect passes from the fertilized egg to the imago through an irreversibly directed metamorphosis, finally forming gametes itself. The fundamental processes of this event in organisms are called "growth and development," by which are understood the most complicated processes—not only the intake of substances but also the differentiation and formation of structures and functions. Moreover, processes are included in a regular manner which lead to the production of new systems of like kind and thus to the reproduction of organisms. These regular changes of state give to the organism a thoroughly historical character, i.e., its present state is essentially dependent upon its previous states and determines its future state. Although this does not distinguish organisms in principle from all other natural objects, yet this historical character is especially clearly seen in them. This rests, on the one hand, on the more or less strong dynamic of the persistent material and energetic exchange of the organism and, on the other hand, on its complex system character, which imposes an irreversible, unidirectional course upon most of the processes concerned. Corresponding to the properties of the organism as an open system, the environment plays a great part in all this as a complex of factors making possible and partly conditioning growth, development, and reproduction. Organisms are normally adapted to their environment in such a way that, in spite of the persistent threat to their existence, a maintenance of the organic forms is guaranteed.

A whole series of the most diverse biological statement complexes deals with the peculiarities of organisms here briefly sketched. What is most striking is the co-ordinated occurrence of temporally ordered processes which lead to the phenomena of growth and development. The statements in question attempt, especially whenever they go beyond pure description, to set up rules, which are empirically testable, according to which predictions can be made about the temporal succession of growth and developmental processes. Such statements can be set up and tested from a great many different standpoints. We frequently speak of the "problem of development" when we have in mind the developmental processes from the fertilized egg to the complete realization of the structural plan, or other growth processes, from the purely morphological point of view. It may here be once more emphasized that, even with such complicated problems, statements of scientific value are possible within the purely morphological point of view. As an example, the allometric growth equation of Huxley and Tessier may be mentioned. It permits predictions about the relative volume and surface changes of body parts or organs in a living system which changes its form through disproportionate growth of its parts. The validity of this law has been demonstrated for a great number of the most diverse transformations of form which an organism can experience during its ontogenesis. For exceptions, supplementary hypotheses were set up of basically the same structure. It is clear that statements like the allometry law are blanket statements under which a whole series of the most diverse and complicated processes is subsumed. The law implicitly assumes that among these processes such a degree of integration exists that a regular blanket statement is not only possible but actually holds good. This feature gives to such statements in biology their special value.

The germ-layer theory may be mentioned as a further example of a predominantly morphologically oriented statement about developmental processes. In contrast to the previous example, we have here a formulation which does not appear in mathematical clothing but works with descriptive concepts and yet, like the allometry statement, is order-analytical in origin. It gives a scheme for the course of the early development of the most diverse animal organisms and for the fundamental processes of organ formation. It also is a blanket statement about very complicated processes of growth and differentiation which are represented in it according to precepts from the morphological viewpoint.

Purely morphological statements about questions of growth and development carry with them the danger of losing their empirical scientific character by slipping into a realism of concepts. If we say that a growth process "obeys the allometric equation," we naturally do not mean that this equation is a "force" which pushes the growth in this direction, and that in this way the manner of growth is satisfactorily "explained." When we see the statements of the germ-layer theory confirmed in the ontogeny of an organism and observe the developmental processes by which the "structural plan of the organism is realized," this, again, does not mean that this structural plan and its various developmental phases are present, as it were, as pre-existent molds into which the "stuff" is poured like a plastic mass. Usually only thoughtless formulations of this kind or such as are chosen in an effort to reach better picturability can lead to serious errors in biological thinking.

Morphological concepts can be defined, in a statement which is useful in empirical science and is free from logical objections, only as assertions about structures which are testable in experience, and they cannot occur as independent concepts in abstraction from this testability. The morphological consideration of processes of growth and development which is applicable to the individual case embraces only visible forms of proportions and spatial relations. But these forms are constructed of living cells, cell aggregates, and their secretory products in the course of complicated processes of multiplication, growth, and differentiation of these living systems. All these processes are also accessible to the other elementary standpoints of biology, and only a synthesis of all possible points of view opens up the prospect of penetrating into these most complex connections. This path has been successfully followed, for example, by experimental embryology. Spemann's theory of organizers is a complex of statements which are empirically testable by experimental interference with embryogeny. They say something about the material actions which issue from certain embryonic regions and about the way in which other embryonic regions react to these actions by growth and differentiation processes. In addition to these, we have statements about the relations between the state of the environment and the developmental processes. The complex interplay of the single processes of embryogeny, governed by organizers of various orders and modifiable by the influence of environment, corresponds with the nature of the organism as a complex open system. The whole physiology of growth and reproduction, the doctrine of hormones and active substances, constitutes a statement complex of a similar kind. In all these statements it is customary to call the magnitudes, measured in experiments, which lie outside the practical surface limits of the organism against the environment the "outer factors," those established within these limits the "inner factors," of growth, of development, etc. But a separation in principle of "outer" and "inner" in such statements is not possible. A hormone which is produced by certain cells of the body is also an "outer" factor for other cells, and the same holds for the action of certain embryonic parts upon the others. Quite frequently, theoretical misunderstandings have arisen through the transformation of the purely conventional significance of the words 'outer' and 'inner' into one of principle.

In a much narrower sense of the word, we can speak of inner factors of development if, turning to the genetical standpoint, we make statements about the part played by the genome as a guiding center of growth and differentiation processes. In an embryological experiment, by operative interference with development, by altering the outer conditions and the like, the present parameters of state of the open system in process of development are changed, and the ensuing changes in the course of development are followed. In a geneticophysiological experiment, by the introduction of a mutated gene, of a genetic defect, and the like, a different state of the whole system is given from the beginning. The typical reaction of the whole system and its parts is altered. Nevertheless, under those changed basic conditions the physiological method of embryological experiment can still be applied. In this way the possibilities of causal-analytical exploration of the processes are significantly enlarged. Statements can be formulated, for example, in which a gene-conditioned active material is represented as an enzyme for the production of certain pigments. Or the moment in development can be demonstrated in which a gene-determined rhythm of certain cell divisions determines the later form of an organ. Or certain complicated transformations in the metabolism are shown to be a chain of particular gene-conditioned enzyme reactions. Statements of this kind connect the concept of the gene or of its alleles, which is empirically testable by means of hereditary analysis or cytogenetic investigations, with morphological and physiological concepts which are verifiable by observation of structures and by physiological and biochemical investigations. They connect structural and functional properties of the genome with structural and functional properties of the organism and are thus statements about the rules correlating two systems of different order with one another. In this borderland between genetics, physiology, and embryology, biological research reaches the highest degree of integration of its technical possibilities but also the highest degree of heuristic fruitfulness.

3. Organic Diversity and Its Structure

The organisms which live on the earth today appear to us in a very great multiplicity of distinguishable types or forms. This multiplicity is a discontinuous one; the types are not connected with one another by all conceivable transitions. This multiplicity is, in fact, very great, but it is finite; we can determine the number of distinguishable types. This work has been completed, at least for certain groups of the animal and plant kingdoms. The given multiplicity does not admit of an ordering by purely quantitative characters, as is the case in, say, Mendeleev's system of the chemical elements. In the domain of organisms, therefore, we could not predict with certainty the properties of a form at present unknown, as was the case prior to the discovery of still unknown elements. Nevertheless, the given multiplicity of organisms is not completely devoid of order, that is, free from any characteristic which permits us to set up a definite order in the multiplicity. On that account men have, from the earliest times, attempted to set up a system of the animal and of the plant kingdom. Attempts at systematic classification according to some one or a few arbitrarily chosen characters have led to the so-called "artificial" systems, e.g., the Linnean system of the flowering plants. We call our present-day system a "natural" system because, in setting it up, as many characters as possible which have proved to be the essential comparable characters, after a careful comparison of the material, are taken into account. The system is based on the concept of species, which will occupy us later and is, in addition, constructed in accordance with the more or less great morphological, possibly also physiological, similarity of the types. Similar species are united to form a genus, similar genera to form a family, and so on. In the establishment of these higher units the systematist proceeds in such a way that he uses for their characterization those properties which strike him, in a comparison of several like species, genera, etc., as special common properties of all members of the groups concerned and which are specially suited to their delimitation from other groups similarly formed. These properties are then distinguished as the generic, family, or order properties. The higher systematic unit becomes in this way an entity which is defined by a few general characters. These are indeed demonstrable—although often in modified form—among the representatives of all species which belong to the higher unit in question; yet the representatives also have the special characters of the species. It is agreed that the higher systematic units, in the first place, are only categories created by the ordering human understanding. Only the doctrine of evolution attempts to see in them natural units, namely, related groups of species with common ancestry. But, quite apart from this theory, it is very peculiar from the purely systematic standpoint that it is possible to unite species into higher systematic units by means of the same ordering principles in all the various types of living things, in spite of their great diversity of organization. In the great multiplicity of living things an order of a very peculiar structure seems to prevail, which is not comparable to the multiplicity of inorganic things.

Every classificatory approach is based on the concept of species. The differences of opinion about the definition and use of this concept in biology are indicative of the great difficulties of this group of problems. The concept of species is used in very different senses by different biologists, according to their point of view and the problem in hand. Nevertheless, most biologists are of the opinion that "the species," in contrast to the higher systematic units, is to be regarded as a *primary* natural unit. The contradiction which seems to occur here is in some degree understandable in view of the present state of biological knowledge but still often leads to misunderstandings. The pure systematist understands by 'species' a type, a "species picture," which is defined by the enumeration of a series of morphological, possibly also physiological and ecological, characters. Organisms found in nature can be recognized as "members of this species" for the most part with certainty by comparison with this type species. It is left to the tact of the specialist—schooled by experience—to regard certain deviatory types as varieties or races within a species or as another species or, in particular cases, as an aberration or anomaly. When we say that "within this genus so and so many species have so far been described," we are using the word 'species' in this purely systematic sense. That this use of the species notion has persisted for the practical purposes of determining individual organisms and assigning them a place in the given multiplicity is for us a proof of the finiteness and discontinuity of this multiplicity and of the constancy—at least relatively—of its ordering structure. In the view of the "species" as an elementary unit of just this order there lies a weighty argument for its estimation as a natural unit. Yet, even for this purely systematic construction of the concept of species, it is necessary in many cases to institute investigations in the field or the laboratory which go beyond the limits of the comparative method. The membership of the very different female forms among butterflies with unisexual polymorphism within one and the same species only becomes clear when their complete fertility with the uniform-appearing males is established. The recognition that drought and rainy-season forms in other butterflies, which look so different, are included in one species takes place only when the natural connection between these generations is ascertained. Here a factor plays a decisive part which is not directly derivable from the comparison of forms—that of the genetic connection.

The "members of a species" usually occupy a continuous habitat in which they form a reproductive community. The requirement of complete fertility within the species is for the most part silently presupposed. Wherever the region of distribution is discontinuous, the virtual fertility between members of a species is taken into account for the delimitation of the concept of species even from the standpoint of the systematist, although not in the first instance. But there are cases in which two groups of organisms, which are distinguishable morphologically either not at all or only with such difficulty that the systematist would at most speak of varieties, are accepted as two species because the two groups of organisms are not fertile with one another and thus form—in spite of a common habitat—two completely separate reproductive communities. On the other hand, there are cases in which species which are well distinguished morphologically, ecologically, and geographically can produce completely fertile hybrids. In yet other cases there are, among the members of a systematically unobjectionably defined species, various groups of individuals which are not fertile with one another within their own group but are fertile with the individuals of another group. This is the effect of genetical sterility factors which coexist in the population. The definition of the species as a reproductive community or at least as a potential reproductive community is thus not quite satisfactory, but it at least denotes a real biological unit.

With the inclusion of the criterion of the reproductive community in the species concept, but especially with the application of this concept in every point of view other than the taxonomic, the concept of species changes its meaning and is to be differently defined accordingly. In such statements, for example, as "This species occupies this or that region, it lives under these particular ecological conditions, it has a particular genic structure," we no longer mean the species type of the systematist but mean instead the totality of the individuals which we customarily include in this species type on the basis of the taxonomic rules and thus a collective of empirically given individual organisms. It is a source of frequent misunderstandings that the word 'species' is used both in the sense of the systematist for the abstraction "species type" and in the other sense for a collective of individuals, but without noticing the diversity of meaning. When we speak of the species as a sum of empirically given individuals, we must specify the point of view according to which this collective is to be marked out and the characters which specify it as a collective of a particular structure. The concept of "species" then passes over into the concept of a "population," which will be discussed in the next chapter. It is impossible within the limits of this monograph to enter into the whole complex of problems connected with the concept of species. This much only is established: that in every biological statement in which the notion of species occurs the definition of species must be so formulated that it is unambiguously explained how the content of the statement is testable in experience.

In the drawing-up of natural systems the systematist works chiefly with the methods of comparative morphology and hence according to order-analytical principles. Experience allows him to choose those statements as the most productive which enable the given multiplicity of forms to be ordered in a way which corresponds best with all biological requirements. The use of the concept of homology, its extension to the developmental stages of animals, the knowledge of the essential significance of certain characters (e.g., of the flower structure in plants) for the setting-up of homologous structural plans, and the like, will cause him to introduce a very diverse valuation of the characters of an organism and to build up the system in accordance with these valuations. The system obtained in this way makes use of the concept of the "relationship" of the various living forms. Two species are said to be nearest related when, within the same genus, they have the greatest number of essential characters in common. In the same sense this concept of relationship is extended to the higher taxonomic units. We are so accustomed today to thinking in evolutionary terms that we often forget that this concept of systematic relationship was originally not meant in the evolutionary sense at all and cannot be used in this sense in pure taxonomy. Without any evolutionary accessory meanings, taxonomy is a scientifically unexceptionable and essential part of biology. It leads to statements about the given organic multiplicity which reveal a series of noteworthy properties of this multiplicity. It is, in the first place, noteworthy that, generally speaking, in the most diverse domains of the organic world groups of species can, according to like ordering principles, be united to form genera and these again to form higher units and that in consequence similar "relationship relations" seem to govern the multiplicity. Further, it is very peculiar that the degree of multiplicity now prevailing in the various domains of the organic world can in no way be predicted according to a general principle or derived from more general laws. While some genera show a great number of species, others are represented by only a single species, and the same holds for the other higher taxonomic units. The potential multiplicity, which we can imagine to ourselves on the basis of what is given in nature as a purely intellectual construction, seems to be realized in the various groups of the world of organisms in very different densities, with very notable gaps or accumulations. Here, also, the multiplicity among living things differs essentially from that in nonliving nature, for the given multiplicity of the elements or of the crystal forms can be derived, more or less satisfactorily, from the more general statements of physics. The purely order-analytical method of taxonomy is not able to undertake such derivations for the multiplicity of living things. The theory of evolution attempts to do this in so far as it transforms the order-analytical statements of taxonomists into causal-analytical ones. In this way the concepts of systematics, for instance, the species concept or the concept of relationship, gain new definitions and new correlations with experience. These new problems will occupy us later (Chap. II, C, 5).

4. The Population as the Natural Form of Existence of Living Beings

In nature there are no species in the systematist's sense but only individuals of various ages, in various stages of development, and in various physiological states, which we speak of as belonging to a particular species—in accordance with the rules of correlation chosen for the definition of the species concept. All together, they form the population, the real form of existence of the species in question at a particular moment of time. Various statements about an organism which are not to hold only for a particular individual and which, moreover, are to be empirically testable in many ways are therefore always statements about the population, about a natural collective. There are various considerations which compel us to formulate collective statements about a population. The mode of adjustment of an organism to its environment; its preservation and successful self-assertion in the environment; the whole complex of ecological, biogeographical, and evolutionary problems requires statements in which not individuals but the natural collective of the population is spoken of and within which the single individuals are distinguished by an average similar norm of reaction. In this way the population is regarded as a system of higher order within which definite relations prevail between the individuals and which, as an open system, stands in mutual relation to the environment. The inner system property of the population becomes especially clear in cases in which the individuals are united into partial systems, accompanied by far-reaching differentiation among themselves, as in the social insects. Also, a united herd as a community of action and a pair in process of reproduction furnish examples of partial systems in this sense. In these and similar cases the whole population is constructed from such more or less firmly united subsystems to form a single supersystem. The population of every organism has its particular structure as a system which can be described by means of a series of empirically testable statements.

The genetical point of view applied to a population likewise reveals its system character. The potential multiplicity of the genome, resulting from the number and spontaneous mutability of the genes, is realized and distributed in various degrees in the reality of a population. The genetics of populations formulates statements about this genetical multiplicity on the basis of experimental analysis. For the distribution of a mutated allele, for the origin of new allele combinations, for the composition of the population out of such combinations and similar processes, the way in which the reproduction and maintenance of the organism concerned are regulated is of the utmost importance. For organisms without sexual reproduction the rules are quite different from those for organisms with obligatory or facultative sexual reproduction. In the case of obligatory autogamous hermaphrodites (for example, many flowering plants) other presuppositions are given than is the case for organisms with separate sexes or even for those the population of which is composed of groups which are sexually isolated from one another. The genetics of populations regards the organism and its population as a genetical system, the peculiarity of which lies not only in its stock of virtual and actual genetic multiplicity but also in the individuality of the apparatus which serves for the distribution, enhancement, or diminution of this multiplicity.

The relations of the genetical system to the environment are conditioned, above all else, by the testing which the gene-conditioned type of reaction gives to the organisms in their struggle for existence. The concept of the selection value of the various gene combinations is defined, in statements belonging to the genetics of populations, not merely speculatively but in accordance with testability by observation and experiment. The study of the population with respect to its potential and actual fertility and its relations to the environment forms the foundation. Statements about the nature of the reproductive relations are therefore of the greatest importance for all population problems. We must take into account whether the reproduction in a population shows panmixia; to what degree the virtual panmixia is actually realized in the population; whether genetic, sexual, ecological, or geographical isolating factors break up the population into several or few separate reproductive communities; and the like.

Ways of Work in Biology

Since statements about populations concern a collective, they very often have a statistical character. The mathematical methods of population statistics form the equipment for all biological statements about populations. Making more or less arbitrary assumptions about the initial magnitudes, population statistics formulates, in a purely mathematical way, the consequences which result from the peculiarity of the organism concerned as a reproductive and as a genetical system. These general statistical statements about the waxing and waning of a population, about the increase and decrease of its genetic multiplicity, about the formation of races, etc., are models which, by the substitution of special magnitudes, become transformed into empirically testable hypotheses in the special, individual case. The whole population of an organism is, in its inner system relations and in its relations to the environment, in an equilibrium, which can experience certain fluctuations and which, by analogy with the state of a single organism, can be called a "flow equilibrium" or a "dynamic equilibrium."

5. The History of Organisms

It is not only the consideration of the single organism which compels us to regard the organism as an open system with a historical character (see Chap. II, C, 2). The momentary state of a whole population is also dependent on its previous states and determinative for the future states of the population, subject to perpetual mutual interaction with the environment. If we extend this point of view to the whole organic world, we come necessarily to the total view that the populations of the various living things stand in mutual relation to one another and to the environment and thus, together with the total environment, form a complex system in which dynamic equilibrium prevails, subject to certain fluctuations and displacements. In such a total view the question of the earlier states of the organic world cannot be separated from the question of the history of the earth, which is regarded as a historical object by the remaining branches of natural science.

The question of the history of organisms would therefore have to be regarded as a genuine problem of empirical science even if we possessed no witnesses concerning earlier states of the world of organisms. But we do possess such witnesses, in the form of fossils which give us a picture, if only an extremely incomplete and defective one, of the diversity of living things which have formerly lived on this earth. The information which these remains give us is especially defective regarding the very long period of the earth's history prior to the Cambrian, during which living things must also have existed. In spite of this defect, paleontology shows us, in a convincing way, *at least* that the multiplicity of organic forms was formerly different from what it is today, that this multiplicity has undergone great changes during the history of the earth, that certain forms of life have died out without leaving any descendants behind and that certain great groups of the plant and animal worlds, which we regard as the most highly organized, have not existed in early epochs. In this way the question of the processes which have taken place in the history of organisms and the investigation of such processes by the methods of empirical science have become essential problems of biology.

The totality of such processes is usually called the "evolution of organisms," and the part of biology concerned with it is called the "doctrine of descent" or "theory of evolution." No subdivision of biology is to such a degree choked up with unrestricted theorizing or fogged by fanciful speculation or made the battleground of extrascientific differences of opinion as this. Especially when the question of the descent of mankind is taken into account, this purely biological problem becomes involved in the arena of philosophical controversies. A clarification of this situation, which is so interesting from the cultural and logical points of view, will be attempted later (Chap. III, C). Here it must be emphatically pointed out that, in the domain of evolution also, only those questions have a legitimate place in scientific biology which are formulated as problems of empirical science. Moreover, the whole complex of evolutionary problems can be treated in biology only in accordance with the principles of empirical science and in this respect is in no way distinguished from other branches of biology. Statements about the history of the organic world have, of course, the peculiarity that they relate in part to epochs of the remote past from which no witnesses of the organisms, or only indirectly accessible ones, have come down to us. For that reason a large part of the statements about evolution will forever have the peculiarity of not being empirically testable in practice and will thus consist of hypotheses about the confirmation of which we cannot judge. If these hypotheses are to maintain their place in science, they must, of course, be so formulated as to be *in principle* empirically testable. They then often have the property that, by analogies with practically testable statements, something can be inferred with a certain probability about the degree of confirmation which these hypotheses might have in case there was a practical possibility of testing them. In no case are hypotheses admissible if their structure is such as to preclude the possibility of empirical test *in principle*.

It is customary to treat the whole complex of evolutionary problems in three groups of statements: (1) Statements which assert that evolutionary processes have, in fact, taken place. Such statements seem to be fully verifiable, in view of the paleontological findings. (2) Statements about the way in which the evolution of the existing multiplicity of living things has, in general or in particular, been brought about—thus the assumption, say, of the gradual transformation of one species into another or the assumption of a phylogenetic tree with dichotomous branching and so of a real derivation of several species from a common ancestral form. In a more comprehensive form, such statements set up a common phylogenetic tree for the whole organic world ("phylogeny"). (3) Statements about the processes in living beings, and between them and the environment, which have taken place during evolution and are going on still and confront us phenomenologically in their totality as the evolution of forms. This group of considerations is often referred to as the question of the "forces, causes, factors, or mechanisms" of evolution. Such formulations conceal within themselves the danger of conceptual realism leading to misinterpretation, as though there were certain scientifically accessible realities, in addition to the processes, in the organism or between this and the environment which "guide" the events of evolution, just as in an obsolete, primitive view of physics the "forces" push and pull the inert particles of matter. For formulating statements belonging to groups 2 and 3 and for their empirical testing, the following possible points of view especially are available: (1) the paleontological; (2) the biogeographical; (3) the comparative anatomical and comparative embryological; and (4) the genetical, that of the genetics of populations and related experimental methods.

Paleontology contains statements about hypothetical evolutionary series, which can be tested by fossil material. When the material lies in easily datable, uninterrupted strata, statements about limited parts of a phylogenetic tree can be regarded as empirically verifiable, although immediate proofs for the natural connection of the generations and the statistical properties of the populations in their full extent are not obtainable from fossil material. Statements about greater phyletic connections and about the whole phylogenetic tree of organisms can be verified or falsified by the material of paleontology only with greater or lesser probability. Nevertheless, paleontology can set up certain more general hypotheses about phylogeny which, being obtained by the method of generalizing induction, can be tested on fossil material. As examples, we may mention the theory of typogenesis and typostasis of Schindewolf and the so-called "Dollo's law." The paleontological point of view is thus quite capable of setting up empirically testable hypotheses about the presumed course of evolution without the help of other possible points of view. Regarding the processes which have taken place in the organisms and between them and the environment during evolution, paleontology naturally has little to contribute. It is misleading when some paleontologists make use of concepts borrowed from other branches of biology—such as the concept of "mutation"—which are scarcely testable on fossil material.

The point of view of plant and animal geography proceeds

from the distribution of the organic multiplicity prevailing at present in its relations to the inhabitable space, and from the displacements in this distribution which are still unambiguously determinable historically. It formulates hypotheses about phyletic connections which are testable by this distribution. The theory of *Rassenkreise* of Rensch and the theory of race and species formation by geographical isolation, as on islands, are examples of the way in which evolutionary hypotheses are formulated as statements which are testable by biogeographical observations. In its retrospective formulation of questions, this point of view finds its connection with paleontology; in its ecological aspect its affinities are with the statistics of populations and the genetic and other experimental methods. Statements obtained by the biogeographical point of view are naturally restricted to fragments of the total event of phylogeny—mostly to the youngest branchings of the phylogenetic tree—while questions concerning its fundamental structure are beyond the grasp of this method. On the other hand, the statements of biogeography and ecology are often connected with the question of the "causes" of evolution, since they deal with the observed distribution of species and races in their testable connection with the environment.

The comparative anatomical and embryological point of view is that from which the idea of evolution chiefly took its origin and from which even at the present day argumentation takes its starting point. The system obtained purely order-analytically is transformed into a causal-analytical one by the introduction of the idea of evolution. The ideal "relationship" according to greater or smaller similarity is reinterpreted as a narrower or broader real blood-relationship. For that reason the statements of systematics and comparative anatomy become more picturable for the thinking human being, and therein, chiefly, lies the charm of this idea. This is all the more intelligible when we consider that the given multiplicity of recent and fossil organic forms cannot otherwise, in its organization, be reduced to any other rules of empirical science and therefore would seem completely unintelligible without the idea of evolution, i.e., would be without evident connection with other data of natural science. Statements about the supposed phylogenetic tree or single developmental series, which we owe to the comparative morphological point of view, still clearly betray their order-analytical origin. The systematist distinguishes his species by means of certain characters and brings together in one genus all those species which have certain common characters, which are now called the "generic" characters. In a similar way the comparative anatomical point of view constructs the hypothetical phylogenetic tree by regarding the special characters of the race or species as the phylogenetically youngest, those of the genus as phylogenetically older, and so on. The hypothetical "common ancestral forms" and "missing links" of the doctrine of descent arise from similar order-analytical thinking, and in this abstract form, in which they issue from the comparative method, they naturally cannot correspond in all points with a real living being. A phylogenetic tree constructed in this way consists at first of structural plans and developmental series of an abstract kind, the correlation of which with experience now must take place on the basis of the hypothesis of evolution, according to rules of correlation other than those used in the purely morphological point of view. For the constructed phylogenetic series the real genetic connection must be demonstrated or made probable in experience or must be refuted. This requirement is often overlooked, and statements about "relationship" or about developmental series in the comparative anatomical sense, obtained by purely order-analytical methods, are formulated as statements about real evolutionary processes; in such cases the idea of evolution enters as a kind of habit of thought which is taken for granted. This holds in a high degree for comparative embryological statements, which, in the form of the so-called "biogenetic basic law," were formerly often very uncritically treated. It must nevertheless be emphasized that all statements obtained by the comparative methods have their legitimate place in biology as hypotheses about the evolutionary process, if they are formulated as statements which are testable in principle, and thus even when they are not testable in practice, perhaps because the fossils can give no information about this part of the process, and the other points of view, such as the bio-geographical and the genetical, also offer no point of attack for a test.

Comparative anatomical research chiefly formulates statements about the supposed course of evolution. Through a functional consideration of the anatomical structures, its theoretical procedure is, however, always connected with the question of the evolutionary factors. The concept of "adaptation" comes from the general experience that organisms in their structures and their functions are so well fitted to their specific environment and are accustomed to react to fluctuations in the state of the environment in such a way that, as a rule, their lives as individuals and as populations are to a sufficient extent guaranteed, in spite of the relatively great threat to their existence. The concept is thus ambiguous from the start: it denotes both a state, that of "being adapted," and also certain processes, that of "adapting one's self," here, in the first instance, in the sense of an individual type of reaction. Yet in the doctrine of descent the concept of adaptation is chiefly used in the sense that in the assumed changes of the species type or in the emergence of new species is seen a process of "adaptation" to changing, or new, complexes of environmental conditions; or this process of adaptation is called the "cause" of the evolutionary process. In such formulations we very frequently encounter, in more or less explicit form, errors which are traceable to conceptual realism, which define "adaptation" as an "active principle" and the change of forms as the "effects thereby brought about." Formulations are only free from objections from the scientific standpoint if regular connections are hypothetically expressed by them between the supposed changes of form and the states of being adapted to the environment, in such a way that they are in principle testable in experience. The attempt to make such formulations picturable very often misleads us into representing evolution as though it had led out of a state of not being adapted, or of being ill adapted, into one of being well adapted. From this there result paradoxical discussions such as that of how, in that case, such a complicated organ as the eye could come into existence by virtue of nothing but small steps of adaptation, when it can only perform its function properly in its completely developed state. But we have no right to assume that there have ever existed nonadapted or ill-adapted organisms, for these would not have been viable. Even the most simply organized living things are ideally adapted to *their* specific environment. Types in which the new adaptation to a changed environment has not been sufficiently successful have obviously died out. In a correct formulation we have to do with the assumption of processes in evolution through which particular adaptations have passed over into other adaptations and thus not properly with adaptation but only with changes in adaptation. The same care in the use of concepts is called for when we speak of "higher development," of "perfecting," or of "progress in evolution." Admittedly, the organization of the higher animals and plants exhibits a higher degree of complication than that of the Protista. Yet this does not mean that the former have acquired a higher degree of adaptation and, with that, of life-efficiency or that we are otherwise justified in introducing a value-judgment. The biological survival value of a species in its proper environment is the only scientifically usable measuring rod.

In dealing with the complex of theories constituting the doctrine of descent, an essential change has come about, since it has become possible, through the progress of genetics, of population genetics, and of related experimental tendencies, to make subdivisions of this field immediately testable either experimentally or by comparative methods. The fundamental assumption of the doctrine of descent presupposes that all those processes which have led during evolution to change in the organic multiplicity in principle also take place today and that the state of this multiplicity at present, as well as every past state at a particular moment of time, is a transition state from a previous to a succeeding one. It must therefore be possible in principle to demonstrate the elements of the evolutionary processes in the organic events of the present day. Those tendencies of research which deal with these questions on the borders of genetics and evolution formulate, for example, experimentally testable statements about the selective value of genetically conditioned differences, about the genetical differences between geographical races, about the genetical differences between species which form fertile hybrids, about the genetical basis of the sexual isolation of races and species, and so on. Comparative genetics formulates statements about the homologizing of the gene stock of related species and about the properties of organisms as genetical systems with regard to the evolution problem. The inference of the genome—of its mutability, its possibilities of new combinations, its structure as a system, and the possibilities of change in its structural basis—opens up undreamed-of new aspects for the treatment of evolutionary questions in the form of scientific, directly testable statements. Questions about the evolution of organisms are often transformed in this way into questions about the evolution of the genome. Admittedly, they do not thereby become simpler, but they become clearer and become accessible in many respects to experimental testing. But, especially, the evolutionary problems become in this way removed from a purely phenomenological study of the types of organisms and become related to elements of which we have some knowledge that they stand as guiding centers behind the phenomena of organic forms and their modes of reaction and that they possess quite definite, unambiguously definable properties. It is therefore not correct to see in the new synthesis of genetics and evolutionary research only a continuation and a more or less important supplementation of the former comparative morphological method. For evolutionary questions, it is far more important when it is demonstrated that in the genome of species there are homologous genes, the properties and function of which in the genome are very well definable and empirically testable by various methods, than when a homologizing of properties is carried out under the morphological point of view. Only those will fail to recognize this who are accustomed to treat everything that can be empirically established as "properties" of a "bearer" existing independently of these properties as a "subject." In this conceptual-realistic view, genes, chromosomal structures, and physiological properties are, just like the peculiarities of form, only "properties" of a species or race, which, from being an abstract entity, becomes the "bearer" of all these properties. Such "species," as well as their supposed phylogenetic changes, are naturally inaccessible to empirical scientific treatment. The individuals or populations whose properties are empirically inferred are scientifically defined only by the totality of these scientifically accessible properties.

Naturally, only questions of race and species formation, of genetic differences between races, species, and genera, are accessible to treatment by the experimental methods just sketched, while these are applicable only to a slight extent or not at all to the question of a genetic connection between the higher systematic groups of organisms. The genetico-experimental point of view restricts its statements, therefore, as a rule to the finer branches of the phylogenetic tree, and leaves open many questions about the total course of phylogeny. On the other hand, it offers the great advantage of contributing important statements about processes during evolution and of subjecting the generally applicable questions about the "factors" and "forces" of evolution to a treatment which is free from objections from the point of view of empirical science. From the point of view of method, we often distinguish today between the realm of "microevolution," which is accessible to genetical methods, and that of "macroevolution," which is less accessible to them, without at the same time wishing to attribute to this distinction too great an importance in principle.

It is well known that for a long time two theories under various forms have played a great part in evolutionary research: the theory of natural selection, sometimes rather inappropriately called "Darwinism," and the theory of direct adaptation, or Lamarckism. Most modern biologists, especially under the impress of the development of genetics, support some variant of the selection theory, often called "Neo-Darwinism." Lamarckism has few supporters today, especially on account of the repeated failure of attempts to verify special hypotheses derived from it. It may be emphasized that it is quite possible to formulate entirely different views about the origin of the given organic multiplicity as empirical scientific hypotheses, e.g., the assumption of an origin of new species from dead substances or from "nothing." Such hypotheses, however, have little prospect of verification by experience and present great difficulties to a consistent incorporation into existing knowledge. The question of the first origin of living things on the earth can likewise only be formulated as an empirical question within scientific biology. It has led to the setting-up of a series of more or less unobjectionably formulated hypotheses, the verification of which has, however, not yet been possible, partly for practical reasons contained in the hypotheses themselves. This question also has been excessively emphasized for extrascientific reasons and made a topic for polemics—often, unfortunately, in a very unobjective form.

Although the whole complex of problems thrown up by the theory of evolution must in many of its parts always remain in the stage of a hypothesis which is not testable in practice, yet it has done invaluable heuristic service in all branches of biology and has therefore become an indispensable part of the method of biology. In view of the multiplicity of points of view regarding evolutionary questions indicated above, it is not surprising that representatives of the various subdivisions of biology, such as systematists, morphologists, paleontologists, biogeographers, and geneticists often put the problems differently and give the hypotheses a different meaning from their several points of view or estimate their empirical confirmation differently. Differences of opinion which come to light in this way have often originated clarifying discussions and so led to fruitful new efforts. Unfortunately, such conflicts are often unpleasant and fruitless, owing to a lack of understanding of the logic of science. In recent years a clear and far-reaching approximation of the various viewpoints has taken place, and various books, as well as discussions at congresses and symposia, allow us to recognize clearly the development of a new synthesis of all possible points of view.

Moreover, in this most difficult branch of biological investigation a phenomenon has become clear which in many other branches of biology, and in all pure empirical sciences, can be regarded as a touchstone for the fundamental confirmation of the methodical path of these sciences: the spontaneous convergence of all lines of development in the science toward a closed, consistent picture of the world.

III. The Significance of Speculation in Biology

Poesis doctrinae tamquam somnium.—FRANCIS BACON

To follow the historical development of the methods and concepts of modern natural history from its prescientific sources and to point out the various phases of this development in its interaction with the contemporary spiritual and social situation would carry us beyond the limits of this monograph. Such an investigation would make much in the present state of biology more understandable than would a criticism purely from the methodological point of view. It would also show that the significance of speculative thinking has played a very different role in the development of this science at different times. Under the indefinite and vague blanket label of "speculation" there are brought together here all possible teachings, doctrines, systems, theories, and statements which deal with living things but which do not satisfy those prerequisites which we have learned to regard as fundamental properties of biological hypotheses and theories. Speculative theories are distinguished from the typical form of statements belonging to empirical science chiefly by the great generality of their sentences, by the imprecise definition of the concepts used, by the lack of unambiguous rules of operation for their use, and by various other features which make the empirical testing of these statements, or the derivation of special testable statements from these general theses, impossible. These general statements are often tautological. They frequently deceive on account of their high degree of "picturability" or by claiming a higher "explanatory value," or they win the reader by seeming to give a simple common solution to very heterogeneous problems of the most diverse fields of natural science.

The situation in biology at the present day is characterized by the fact that the scientific drive of its various disciplines is almost exclusively oriented in the empirical direction and scarcely takes note of the existence of speculative movements. In the textbooks and manuals, in the numerous periodicals well known to the specialist, we shall look almost in vain for essays of a speculative kind. At the great international conferences in which the latest advances in the biological disciplines are reported and in which, by lively exchange of ideas, the directions for further work are obtained, and in narrowly limited symposia serving to synthesize various research tendencies, speculative ideas are scarcely mentioned. And yet there is a large genus of literature of this kind which is distributed in the form of corpulent books, of brochures, of discussions in philosophical journals, of reviews of philosophical congresses, and of articles in general or popular scientific magazines. They are rarely read by experts and are absent from the specialized libraries, and yet many books having a speculative biological content reach higher sales than the best textbooks. Their readers are chiefly educated laymen, the representatives of neighboring sciences seeking a general orientation, and persons of general education. Through the reading of such works the erroneous opinion is often propagated that "biology" is a peculiar, predominantly speculative science which is antithetical to the empirical sciences of zoölogy, botany, etc., and even that these are subordinate to it. Among the authors of the speculative literature are some philosophical experts who have themselves never worked at biology and often have only a very imperfect and superficial understanding of the biological literature which they have read. But we also often find among them biological experts who have acquired great merit by research but have abandoned it in their later years and devoted themselves wholly to speculation. Another group consists of prominent representatives of biological disciplines, frequently of those belonging to applied biology—

doctors, technicians, and the like—who in their busy lives have found few opportunities for occupying themselves with fundamental scientific questions but occasionally feel themselves obliged to publish speculative effusions. To follow the psychological roots of these phenomena would be an attractive task but one which lies quite outside the limits of this work.

It is in any case a phenomenon which is so clearly impressed on no other branch of natural science and is especially characteristic of biology that alongside the system of expert science such a rich speculative literature should grow up, without there being at the present day a living connection worth mentioning between the two fields. There are a few leading biologists who have undertaken the troublesome task of a critical analysis of speculation from the empiricist standpoint: among them are M. Hartmann, A. Bünning, H. Winterstein, and W. Zimmermann. It is certainly no unjustified reproach when the theorizing authors complain of the complete lack of interest in speculative ideas on the part of most scientific biologists. Much leading-astray by worthless literature, much conscious or unconscious misuse of science, much baseless spiritual conflict, could have been avoided if more interest in general questions had been shown among empiricists. Many platitudes and crudities of the empirical specialists, many errors of the universalist striving after knowledge, could be done away with, if, in the course of a lively discussion based on considerations from the logic of science, it could be shown more clearly what the tasks and limits of empirical scientific thought and the prospects for a synthesis with other intellectual possibilities are, especially in the field of biology. In this positive sense the following attempt at a short survey should be judged, in spite of its partly negative criticism.

A. The Psychological Function of Speculation in Scientific Biology

It is not to be denied that speculative ideas of a general kind, which cannot be regarded as empirical hypotheses in the proper sense, can have a legitimate function in biology by promoting the development of this science. There are even some authors who see a danger to the future development of modern biology in its impoverishment of speculative inquiries. It is without doubt the case that in the field of biology, as well as in other fields, the present-day empirical structure of science has gradually developed from a state allied to the philosophy of nature of former times, whereby the speculative views of prescientific and early scientific times have formed the historical, spiritual background. After the European Middle Ages had for the last time offered a world picture of astonishing unity and completeness, we witnessed a gradual dissociation between empirical research and the philosophy of nature. After a period of empirical progress in the sixteenth and seventeenth centuries, there followed in the eighteenth century a time of stronger emphasis on ideas belonging to the philosophy of nature, which exerted their influence far into the nineteenth century and in which are to be sought the roots of many concepts of modern natural science. These concepts have, of course, strongly changed their meaning in the course of development, and many problems and alternatives of that time have become pointless through the progress of empirical scientific work. Nevertheless, it was these concepts that stimulated the research and first gave it direction.

An example of this is furnished by the opposed views of preformation and epigenesis in the ontogeny of organisms, which in various transformations are to be recognized throughout the history of biology. Both have their roots in primitive, prescientific ideas. While the philosophical views of antiquity and of the Aristotelian medieval Scholastics seem rather to have supported the ideas of preformation, the Christian philosophy of Augustinian origin preferred epigenesis. In the eighteenth and nineteenth centuries the two views stood opposed to each other in the form of fundamental biological theories. The present state of knowledge has shown this disjunction to be pointless. These discussions between the two tendencies, carried out chiefly in the field of the philosophy of nature, were often helpful for empirical research by giving it direction but were also sometimes cramping and misleading. In any case, the historical roots of the concepts of modern genetics are often traceable to the theory of preformation, while the concepts of developmental physiology often have their origin in epigenetic views. The general theory of sexuality of M. Hartmann provides an example from recent times. The author of this monograph has attempted to show that the general statements of this theory are tautologous and do not have the properties of unambiguously defined and heuristically useful hypotheses (1933). Nevertheless, the theory of sexuality of Hartmann has given a stimulus to a great number of valuable experimental investigations which represent an essential enrichment of our knowledge. The special hypothetical starting points of these investigations come, of course, in all cases from the complex of theories of stimulus physiology, developmental physiology, biochemistry, and genetics, and the results accordingly fit into these fields without either proving or disproving the general statements of the theory of sexuality. Their interpretation by means of the ideas of this theory seems only a superfluous decoration. The function of such speculative theories of a general kind obviously consists only in forming a conceptual background offering incentives of a general kind, revealing possibilities to thought, and in this way enlivening the investigation, while the setting-up of working hypotheses always arises from the current state of the science in the form described in the first chapters of this work. The importance of speculation seems to decrease more and more with the progressive increase in knowledge and the perfecting of empirical methods. The author does not venture an opinion on the question whether this process necessarily takes place according to a law of intellectual development.

In this connection the question arises whether there exists a "theoretical biology" as an independent science. Some authors, especially von Bertalanffy have declared themselves firmly in favor of taking this branch into consideration in the organization of teaching and research and have drawn attention to the parallel case of theoretical physics. But I believe that this is not a correct comparison. In the case of physics the mastery of a specific apparatus of applied mathematics presupposes special gifts and special methods. Yet even here the boundaries between theoretical physics and experimental physics have become very much blurred in recent years and were in part only set up through the misunderstanding of certain physicists about the development of their science. In the field of biology the situation is quite different, inasmuch as here at present no special mathematical apparatus is necessary for the setting-up of a system of theories but, on the contrary, an enduring contact with experimental research. Even the biophysical ideas of Rashevsky, constructed with a great display of mathematics and abstraction, or the apparatus of biostatistics and the mathematical methods of genetics maintain their significance only in continuous connection with experimental research. A deliberate separation of a "theoretical biology" would today mean an intellectual decline or even the encouragement of speculative tendencies which would not promote the development of the science. A *purely* theoretical biology would be unable to make any scientific assertion which would say more than the statements of the special branches about living things or to which the latter would be subordinate. The concept of "general biology," as it is used in the present work, means nothing more than a synthesis of the single biological disciplines. In this sense it is certainly justified from the teaching point of view but not as a branch of research, since specialization is not then admissible. With the increase of knowledge the integration of the results of the special branches of biology becomes an ever greater, but at the same time a more and more necessary, task. It will best be solved by the working together of experimental biologists who possess the endowments necessary for the task. But the demand for the establishment of a special field of research of "general biology" with its own characteristic methods is not in the least justified.

B. *Parabiology: The Misuse of Speculation*

Under this heading are gathered together speculative tendencies, systems, and statements which in my view either do not exercise at all the legitimate function of speculation described in the previous section or do so only in part or are inclined to surpass the boundaries of this function. This designation is not intended to have a derogatory sense but only to indicate that these tendencies stand "alongside" scientific biology, without showing any scientifically unobjectionable connection with it. That is not to say that these tendencies cannot have other, perhaps very valuable, functions in the life and thought of mankind. For the progress of biological science, however, they seem to be useless, or even harmful in so far as they mislead non-biologists about the real character of scientific biology and divert the expert from a correct formulation of his problems.

Many fundamental misunderstandings owe their origin to the ambiguity of the word 'life,' by which we denote not only the state of living natural objects as it can be dealt with in empirical science but also the subjective recognition of our existence—and thus introduce a problem of quite another kind, which has nothing to do with the scientific tasks of biology. Even for the expert it is not always easy to avoid these overtones of meaning of the word 'life' when he wishes to use it only in the sense of his science. Connected with this is also the fact that in many parabiological systems an open or concealed conceptual realism is involved in using the word 'life.' In popular expositions we often read: "Whatever new knowledge biological science may have brought to light, what life really is, what its nature may be, we know just as little as formerly; it is, moreover, an unsolved an insoluble problem." In such and similar contexts the concept "life" is personified into a natural object existing independently alongside the living organism. In this sense "biology" then becomes the proper science of this "life," separated from the biology of scientific experts, or having the latter subordinate to it. Some other concepts of biology are also involved in a similar conceptual realism, such as "form," "configuration" ("Gestalt"), "wholeness," "species," "type." That no scientifically useful beginning can be made with such ideas, that they must, on the contrary, always remain sterile speculations, without connection with scientific biology, is indeed self-evident. In what follows, a short survey will be given of various systems of a parabiological kind, without attempting a complete or exhaustive treatment.

The Significance of Speculation in Biology

1. Mechanism and Vitalism

The alleged antithesis between systems of mechanism and
systems of vitalism is frequently regarded as a very important
and fundamental problem of biology. These two views have appeared in various special forms and modifications and today
still have their representatives here and there among scientific
biologists. The majority of experts take, of course, a rather de-
tached attitude toward them. On the contrary, these views re-
ceive lively discussion in circles interested in the philosophy of
nature, in which, today, various kinds of neovitalism find supporters. Regarded from the empiricist standpoint, the mechanistic and vitalistic systems show a fundamental affinity, while the
contrasts between them seem to lie more in the formulation.
Common to both systems, in the first place, is the view of
causality in the sense of an executive causality, while in the empirical science of the present day the causal connection is usually
conceived only in the sense of a consecutive causality. Both systems seek for an "explanation" of the organic event by reducing
it to a "principle" or a "category" which is "at work" in it,
which "causes" it. The concepts of "principle," "category,"
etc., are reified into "working causes." It is of secondary importance whether the mechanist locates these causes in definite
hypothetical "elementary vital units" or in the fundamental
properties of "matter" or whether the vitalist seeks them in an
"entelechy." There have, indeed, been remarkable combined
systems, as, for example, Reinke's theory of "dominants" or the
"biodynamic guiding fields" or Pflüger's theory of "life-stuffs"
which veil themselves in a physicochemical garb but are essentially vitalistic. On the other hand, H. Driesch has called himself, not without justice, a "consistent mechanist" because his
theory presupposes a machine model of the organism and an extreme preformationist view, in which the entelechy "is the en-
gineer who sets this machine in motion."

A further common feature of mechanistic and vitalistic systems is the tautological character of their general statements.
When the mechanist ascribes properties to his "biomolecules,"
when he assumes structures in the fertilized egg from which all

vital and developmental processes necessarily follow, this amounts to no more than a pure tautology, in which what is to be explained is already put into the definition of the concepts serving for explanation. The same is done by the vitalist with the entelechy, which is always required to do what is expected of it for the explanation of the vital and developmental processes. In the statements in which the concept of entelechy or something similar occurs, these concepts are, for the most part, defined only by their connection with the processes which they are to cause. In this way such statements escape empirical testing either by experiment or by observation. It is for this reason that the formulations of mechanism and vitalism have been discussed for many decades on the purely speculative plane, without the great progress meanwhile reached in research having permitted a decision in this controversy. Every new result can be interpreted in the expressions of mechanism or in those of vitalism. There is no *experimentum crucis* which can decide between the opposed standpoints. Naturally, attempts have not been wanting which sought to formulate the statements of the great speculative systems in a manner which is empirically testable or to derive special formulations from the general statements which are empirically testable. The famous examples of the older vitalism—for instance, the experimental separation of cells during cleavage stages—were admittedly proofs against special hypotheses derived from preformationist ideas, such as Weismann's theory of the "genetically unequal" cleavage division, but are not proofs for or against the general theorems of vitalism. General assertions about the "entelechy" or the "forces immanent in the system," for example, avoid giving any information about the difference between the mosaic and the regulative type of cleavage; they cheerfully neglect the question of why the regulative ability has limits and where they are. If from these general statements we try to derive special ones which are testable in experience, then this can only be done by means of the transition to hypothesis formation on the basis of the existing state of empirical research, in accordance with the rules sketched in the earlier chapters of this work. The researches of

Spemann, Hörstadius, Plough, and others have shown, that special, more or less narrow limits are set to the equipotential system and that these limits can be defined by causal analysis of the conditions given in the system or in the environment which are empirically testable. Development takes an atypical course just as consistently and in a manner just as much open to experimental analysis as it does a typical one.

It is characteristic of the newer vitalism that its supporters present a special, perhaps intentional, lack of understanding of the results of genetics and their connection with the achievements of developmental physiology—for example, Woltereck (1931) and Driesch (1944). The classical formulations of the great mechanism-versus-vitalism antithesis are adapted to the state of biology which prevailed at the time of its origin. Like all speculative systems which strive after "final ontological" statements, they cling to the current state of empirical science, which they attempt to petrify in its main features. The results of genetics are therefore as far as possible represented as in principle unimportant or are reinterpreted into "problems of higher order." An idea of genetical developmental physiology—the temporal co-ordination of gene or enzyme action—is emphasized as the problem of "insertion," and this is ascribed to the intervention of the entelechy. Such an assumption could only hold as a true scientific hypothesis if the laws of the relations between the entelechy and its effects were empirically testable by an appropriate definition of the concepts used, under varying experimental conditions. But as soon as we attempt such a formulation, the vitalistic notion of entelechy disappears and gives way to the empirical determinations of the magnitudes of the system under consideration. Driesch, the most consistent thinker among the vitalists, therefore tried in his later works to remove the concept of entelechy more and more from the grasp of empirical scientific methods. While some vitalists would see in the entelechy something "in principle analogous to the forces of inorganic nature," Driesch regarded it as "not energy, not force, not intensity, and not constant" but as something "immaterial, nonspatial." It is not so much this vague definition which makes the central concept of vitalism scientifically unusable as, rather, the complete lack of hypothesis construction which could contribute to the enlargement of the science by empirical testing. The same is true of the materialistically formulated general basic concepts of mechanistic systems.

A feature common to many parabiological systems is the more or less anthropomorphic character of their formulations, on which, in part, their seemingly explanatory value and their effect on the public depend. As one of many examples, we may mention the considerations which Ballauff (1940) connects with the establishment of structures or processes of a complex kind in the organic world. He holds that in such cases we always speak of "order," but we must always ask where the "orderer" is. A point of view which for every order requires an orderer and for every "law" a lawgiver—some neovitalists say that the role of the entelechy is not "executive" but "legislative"—is typically anthropomorphic, and attempts have been made to describe the vital processes and the behavior of all organisms, right down to the unicellulars, in the language of human psychology. Such an attempt would be perfectly admissible methodologically if only we had the possibility of making statements about the psychic life of organisms which are testable by inner experience in the same way in which statements about the conscious processes of human beings are testable. For that reason extreme psychovitalism is the only form of vitalism which has been thought out thoroughly. That such a procedure is for many reasons not open to biology is clear to everyone, and therefore also the concealed forms of the anthropomorphic point of view must be rejected as inadmissible.

Biologists of the empiricist persuasion are usually called "mechanists" by vitalists. It is objected against them that they attempt to reduce all organic events to the simple physicochemical laws of inorganic nature, which is held to be impossible. What we are to understand by "reduce" and "the simple physicochemical laws" of inorganic nature is, of course, quite inadequately defined. The scientific biologist is far from beginning work with the preconceived opinion of such a reducibility.

What matters to him is that he should use those exact empirical methods which have proved themselves to be successful in his science. *In principle* these are, of course, the same as those used in the inorganic natural sciences. No biologist will assert that the results of the physics and chemistry of inorganic nature will alone suffice to elucidate scientifically the relations of living things. In organisms we find materials united in the most complex metabolic relations with one another and with the environment in the form of highly complex open systems of peculiar hierarchically ordered structure, such as we do not find in non-living nature. Without the knowledge of living things we should never have had examples of such structures, and an essential part of what is given in nature would have remained unknown to us. The laborious scientific exploration of this field has therefore also given rise to sciences with independent special methods within the boundaries of natural science as a whole. Nevertheless, no man doubts that even within organisms the same general physicochemical laws hold as in inorganic nature. The elements show the same reactive properties; the same general kinetic rules hold, and the same thermodynamic relations. Moreover, the often-cited apparent contradiction to the second law of thermodynamics holds only if we falsely regard organisms as closed systems. But it vanishes as soon as we consider the organism with its environment and thus a larger section of nature.

An argument often brought forward from the side of a philosophy of nature is the seeming "improbability" of organisms. We cannot here discuss the ambiguous use of the notion of probability by many authors. In so far as this objection relates to the structure of living things and their energy relations, the matter has been made so clear in recent years from the physiological side (Bünning, 1949) and from the physical side (Schrödinger, 1944) that a general discussion of it in vague terms should be superfluous. Structures which are improbable when considered from the point of view of inorganic models are formed by the special biochemical and biophysical situation in the organism and are thus very probable. The peculiarity of organic events does not lie in peculiar "forces" which do not occur in inorganic nature but in the exceedingly complicated systematic connection of the materials in their interactions. In view of this organization, the energy changes in the organism are no less "probable" than every other physical occurrence; on the contrary, they are the only probable ones. Peculiar relations prevail in the parts of nature represented by organisms just in so far as the high degree of order (or the thermodynamic potential) becomes greater by an increasing disorder in the rest of nature. Such relations would scarcely be conceivable or predictable from a knowledge of inorganic nature alone. Nevertheless, the experience of inorganic and that of organic science join together to give a unitary and consistent picture of the world, derived from the fundamental unity of method. This characterization of the present-day situation in biology does not signify mechanism in the above sense, still less the acknowledgment of a philosophical doctrine in the sense of materialism.

The attitude taken by biologists toward the great speculative systems is often characterized by some uncertainty and too much aloofness (e.g., Spemann, 1936). There is no point in contesting the doctrines of vitalists and mechanists by factual arguments, as M. Hartmann and L. von Bertalanffy do in so praiseworthy a manner, if the essential point is not emphasized, namely, the methodological defects of these systems. For that reason it is a mistake to adopt the standpoint that, although in the present state of investigation a decision between the alternatives presented by parabiological ideas is not possible, yet such a decision is to be expected of some future state of biology. Empirical science will never make possible a decision about questions which are formulated in a way which excludes treatment by empirical methods. From the point of view of method it is also very precarious to use the concepts of vitalism in order to denote the boundary of the unexplored or unexplorable in an indeterminate manner, without trying to seek a purely epistemological clarification of the question of such a boundary. Again, other biologists are of the opinion that the mechanism-vitalism antithesis is to be overcome by a synthesis to which both terms are subordinate; yet such attempts easily lead to new formulations of a parabiological kind. Finally, there is a series of speculative theories which, although not openly basing themselves on the systems mentioned, nevertheless have essential features in common with them and, in spite of their often cautious and careful formulations, have a parabiological character. Some of these will be discussed in the following sections.

2. Subjectivity, Activity, Purposiveness, and Conformity to Plan in the Organic World

Some authors believe that living beings are essentially distinguished from inorganic things by their "subjectivity." The notion of "individual" is often linked with this idea. It is clear that the roots of this view lie in the personal experience of human beings, who feel themselves to be indivisible subjects. Closely connected with these ideas is the assertion that organisms are distinguished from the rest of nature by a special "activity." The difficulty of using the notion of "individual" empirically in an unobjectionable way in a physiological view of organisms has already been discussed. According to the level of organization and other circumstances, an "individual" is divisible by the formation of complete new "individuals," or it can regenerate essential parts after loss or support such loss without regeneration. The notion of "individual" is not used in biology in the original sense of the word but only conventionally and must be specially defined for the special case. If this is done, then the notion of "subjectivity" also disappears from physiological analysis, unless we wish to withhold it, as a vitalistic fiction, from such analysis. The same holds for the "activity" of the organism, in so far as more is meant than a statement complex about the energy exchanges between organism and environment. That the study of behavior and other branches of biology must often use these and similar expressions in its statements in order to make them picturable is self-evident. It will not be difficult for the methodologically educated to estimate the borrowed expressions correctly, serving as these do the interests of picturability, without attaching an ontological significance to the concepts used in them. This holds especially for the distinction made in various connections between "active" or "living" and "passive" or "dead" parts in organisms or cells. Here, also, such a distinction serves only to illustrate certain physiological facts but has no ontological significance. Analysis always leads only to complicated reciprocal actions of the equally indispensable constituents all contributing to the construction of a highly organized system, but without a special degree of "vitality" or "activity" being ascribed to any one of these materials or structures. It would be a mistake to seek in the organism for special "living" structures, for "centers of activity" or an "active principle" of any kind which makes use of the remaining materials and structures as only the passive "substrate" of its action, as is done in principle in all mechanistic and vitalistic systems. The same holds for such expressions—very often used in biology—as 'potency,' 'primordium,' 'capacity,' 'factor,' 'tendency,' etc. In genuine scientific statements these words never denote independent powers concealed in organisms from which some special activity proceeds but are only illustrative circumlocutions for the assertion of a law or for the definition of a set of conditions which are testable in experience. Although we are again and again compelled by the high degree of complication of organic events deliberately to simplify by considering single causal connections outside their relation to the system, and although it is advisable to use language illustratively in the exposition of such connections, still we must always be aware of the danger of misunderstanding which this involves. Such commonly used notions of scientific biology as "gene," "enzyme," "hormone," "stimulus," "excitation," and the like, are also never quite safe from such misinterpretations and are often used, especially in popular expositions, in a parabiological sense without the author intending, or even being aware of, the fact. It is perhaps more difficult in biology than in the other natural sciences to be popular without being misleading.

The frequently used expressions 'purposiveness' and 'conformity to plan' are also dangerously ambiguous. When we describe the structures of an organism and their functions within the system and in reciprocal action with the environment, we are naturally at liberty to call all these arrangements and functions "purposeful" in so far as the ability to live and the peculiarity of the system depend on just these arrangements. A system having unpurposeful arrangements is simply not possible under the given circumstances. We can with equal justification ascribe "purposefulness" to every inorganic thing if we regard being as it is as its purpose. It is clear that such a highly complicated open system as a living organism, whose existence is so strongly threatened, must possess a great number of arrangements in order to preserve itself. The fact that in spite of these arrangements a very great number of living things always perish for external reasons, that in the case of certain types of organization every organism perishes at a certain age for internal reasons, shows how difficult it is to preserve these complicated open systems in nature. Moreover, many species of animals and plants can be shown to have died out in the course of the earth's history. The organizations of organisms thus prove to be successful only under particular constellations of external and internal conditions and are purposeful in this sense for the maintenance of the individual and the species. The significance and origin of a purposeful arrangement, such as the elaborately constructed mouth parts of an insect, can be depicted scientifically only from the following three points of view and thereby "explained": (1) its structure and its function and their significance for the preservation of the life of the animal are described; (2) the genetical and embryological conditions of its origin in ontogeny are investigated; (3) as far as this is possible, the processes in the phylogenetical origin of organisms with such arrangements are investigated. The employment of the concept of purposefulness in biology in this sense is perfectly correct. When the question of the function of an organ and of its significance for the survival of the individual or the species is called the question of its "purpose," then this kind of teleological point of view is quite common in biology and has high heuristic value. But it does not *fundamentally* distinguish the procedure of biology from that of other sciences of nature. For even in nonliving systems it is possible to ask about the significance of the single constituents for the preservation of the system. That it plays such a significantly greater part in biology and is so uncommonly successful heuristically rests only on the fact that living systems exhibit such a high degree of complication. It is part of the definition of a highly complex system that each of its structures and functions—allowing for variations within limits—must be necessary for the maintenance of the system in its particularity and is in this sense purposeful. But in many speculations of a teleological tendency the concept of purposefulness is used, more or less disguised, in that other sense which it is given in everyday language. Here arrangements or actions are called "purposeful" if their aim is imagined by human beings as a goal which is striven after. Extreme finalists seek to "explain" all organic events, even processes of phylogenetic and ontogenetic development, by a "striving for a goal," which they see as a special "principle" which is "at work" in organic things. As in other vitalistic concepts, here also a consistent psychologizing, according to the pattern of human conscious processes, provides the only possibility of making such formulations testable in experience. Later we shall discuss the noteworthy attitude taken by finalists toward the notion of causation.

The situation is similar in the case of certain general statements about the "conformity to plan" or "planned order" in organic beings. A general statement about a persistent order easily becomes tautologous if special testable statements about the rules of this order are not derivable from it. Also, the order-analytical statements of biology, obtained by purely comparative means, possess scientific value only through their quite special content which is testable in experience. Statements about the degree or the height of an order, also, are only useful when they provide a means for measuring this degree. The same holds for the concepts "plan," "constructional plan," "functional pattern," and the like. It is characteristic of many parabiological theories that they turn such concepts into things to which they attribute an action on the "substrate" of organic events. In this way some philosophers of nature speak of an "antecedence" of the structural plan in the developmental processes in ontogeny. The apparent explanatory value of such ideas lies wholly in the purely anthropomorphic conception which is behind them.

It is not to be denied that the teleological mode of expression makes biological topics very clear and is therefore almost indispensable, especially in teaching biological subjects. Nevertheless, it is doubtful whether the statement "So long as an organic system has not yet reached its greatest possible development, it strives toward it" is to be regarded as a "general principle" of organic development. If such a sentence is not intended in an extreme psychovitalistic sense, it says nothing at all.

3. Wholeness and Causality

In many recent philosophical systems the concept of "wholeness" (*Ganzheit*) plays a central part. Its fundamental assertion is that "the whole is more than the sum of its parts." By many authors the peculiarity of being a whole is regarded as especially characteristic of living things, where types of wholeness of still higher degree are distinguished, e.g., species, biocoenoses, peoples, races, etc. Other authors allow nonliving objects such as atoms also to have this property of wholeness. The principal difficulty for a critical evaluation of such statements lies chiefly in the fact that in them the concepts "the whole," "the parts," and "the sum" are not unambiguously enough defined to be useful in special testable hypotheses. In some of these statements "the whole" signifies no more than our idea of a living thing which we consciously experience as a closed unity, and "the parts" are those ideas of parts into which we divide this whole conceptually according to taste. In such statements the question of the relation between the whole and its parts is a logical or psychological, but not a biological, one. In so far as it is a question of empirically comprehensible parts, or part-processes, in an organism, it is already laid down, in the more or less satisfactory definition of these parts, in what regular relations they stand to other parts of the whole. The word 'sum' in a purely mathematical sense is then never sufficient for a characterization of these relations. Even where we speak, for example, of a summative action of individual processes, a derivative and deliberately simplified mode of expression is being used. Physiological analysis leads us again and again to the complicated mutual actions in which all the constituent parts of a living being are systematically involved with one another and with the environment. It would be a mistake to expect that the processes in an organism could be understood from the sum of all those reactions which its parts would perform if they were not in the union of the living system, for then they would be under quite different sets of conditions and would therefore react quite differently. If the above-stated principle says no more than that in organisms such complicated systems of conditions always prevail, then it is quite justified. Then the notion of "wholeness" is only a paraphrase of the concept of system and is also applicable to systems of higher order, such as species and biocoenoses, as well as to inorganic systems. The same holds for another frequently repeated formulation, namely, that "all processes in an organism are related to the whole." There is nothing objectionable about this if it is only an attempt to characterize the type of a complicated system. But it is necessary, if error is to be avoided, also to take into account the relationship to the environment. In a more special sense the "purposefulness" of the partial processes in an organism is also intended by this statement, i.e., the partial processes run their course in accordance with the maintenance of the whole system (see Chap. II).

Many philosophers of nature wish, of course, to give to the concept of wholeness an explanatory value extending far beyond these limits. This is expressed, for example, in the formulation "The whole determines the parts, not the parts the whole" or "The whole is the cause of the particular course taken by the single processes." Such assertions presuppose that there is another path to the immediate comprehension and definition of the "whole," as well as the analysis of all its structures and partial processes, than the methods of empirical science available to biologists. In making use of the Gestalt experience which we have in considering a living thing, the morphologist is especially prone to fall into a realism of ideas in the sense of an "intuition of wholeness" or an "experience of a formed totality." It is an age-old need of human beings in considering nature to raise the deep emotional impression which nature and her creatures make upon us into an explanatory principle. Thus we often try to see, behind the forms of living things, especially of the human body, laws which derive their meaning from purely mathematical considerations, aesthetic sensations, or ethical feelings and thus from regions of the inner experience of mankind. The theory of proportions of Albrecht Dürer is one of many examples. But such hypotheses are useful in empirical science only when their elements are testable in outer experience. The concept of "Gestalt" as an immediately comprehended totality has its legitimate place, therefore, only in statements belonging to the study of behavior, to the physiology of the senses, or to experimental psychology, in which the reactions of living things are dealt with. For the more highly organized living things react to all the stimulations in their environment *as if* this were a "Gestalt." Thus, in the statements mentioned, something is expressed about the relations between a particular environment and a particular mode of reaction of the organism which can be tested empirically, with the help of the hypothetical concept "Gestalt." This has nothing to do, however, with the question of in what manner the forms and structures of living things are scientifically accessible in external experience. An idealistic morphology, which tries to adopt toward biology somewhat the same relation that geometry has to physics, will always exhaust itself in tautological assertions. Zimmermann (1938) has called the lack of subject/object separation the fundamental error of this way of thinking and has discussed the arguments against a fundamentally different procedure in the treatment of morphological and physiological questions.

In connection with the discussions of this and of the previous chapter, reference should be made to the various views about the notion of cause which prevail in speculative systems (see also P. Frank, 1932). The teleologists distinguish between "efficient causes" and "final causes," in which they regard "final causality" (in the sense of aim or purpose causality) as an essential characteristic of fundamental biological processes. A statement formulated according to empirical science—"The state B follows regularly on the state A"—is customarily interpreted in such a way in ordinary language that A is called the "cause" of B. Nevertheless, nothing at all is altered in the statement and its empirical testability if we call B the "cause" of A. Only in the case of human actions is it otherwise, in so far as the state B, intended by the acting human being, is already consciously present as an aimed-at goal; and it is the knowledge of this connection which permits us so to transform the above statement that we include this fact in the definition of the state A. Thus the attempt to make a distinction between efficient and final causes seems meaningless without an anthropomorphic psychologizing of natural processes. Idealistic morphology also seeks to introduce, in its idea of the "archetype" and its relations to development in ontogeny and phylogeny, a kind of final causality into the morphological point of view (cf. Troll, 1948). If rules which are to be testable in experience are set up for the occurrence of an ontogenetic or phylogenetic developmental process, it is again quite irrelevant whether we wish to call the initial state the "cause" of the final state or, vice versa, to call the "form" reached in development the "cause" of the developmental processes. Only in the case of an inadmissible reification of ideas into natural objects and an anthropomorphic reinterpretation of the goal or aim concept does the so-called "archetype" thinking of the morphologist seem to be different from the so-called "causal" thinking of the physiologist. But then the formulation at once loses its usefulness for empirical science. In close connection with the mode of procedure of physiology some authors have believed that they could recognize a special form of causality in "impulse causality" or "catalytical causality" which ostensibly occurs only in certain processes in organisms, for example, processes in stimulus physiology or processes governed by enzymes (cf. Mittasch, 1938). But the peculiarity of the causal process in such occurrences persists only so long as we isolate partial causes from the event and do not take fully into account, or do not fully know, the constellations of conditions given in the system itself. It is here also, as with the processes supposedly to be explained only by "wholeness," that empirical science makes use of just such effects, which seem inexplicable by the causal connections hitherto known, in order to discover new parts and new mutual relations in organisms. The use of the methods which have proved themselves in science always leads to a consistent synthesis with the existing knowledge and to an enlargement of the system.

Some developments of speculative biology have concerned themselves especially with the question of the indeterminateness of the processes in organisms. It is clear that many biological laws can be only statistical laws in the sense of classical statistics. These are all statements about collectives and collective processes, like the Mendelian rules of inheritance. Almost all biological processes take place in the macrophysical domain, so that in their case the indeterminateness of the microphysical elementary processes plays no part. This also holds for enzymatic, hormonal, and stimulus-physiological processes, for which an origin from the microphysical domain has often been incorrectly asserted on the basis of superficial estimates. The single exception is the very important process of mutation. The role of the mutation frequency within a population or of the somatic mutative occurrences within a body is therefore only statistically describable. But the action of a mutation when once it has appeared in the life of a cell or of an organism and in their descendants again lies entirely in the macrophysical domain. The idea of P. Jordan of an amplifying mechanism which could allow the processes in the microphysical domain to influence the macrophysical organic event is quite conceivable. But we know no such phenomenon in the behavior of organisms, and it would be difficult to imagine how such effects within an organism, which are only statistically predictable, can be reconciled with its existence as a highly complicated system. The conceptually quite mistaken relation into which Jordan has brought his "acausal amplifier theory" with the question of the human freedom of will has created for the whole theory an undeserved popularity. Meanwhile, clarification from various directions (e.g., Hartmann, 1948, and Bünning, 1935) renders a discussion of it unnecessary. On the clear conceptual distinction between the fundamental quantum-theoretical indeterminacy (e.g., of a single event of mutation) and the classical statistical indeterminacy (e.g., of the processes of selection) and on the correct evaluation of the great importance of such events for processes on the phylogenetic plane, especially clear ideas have been expressed by Möglich, Rompe, and Timoféeff-Ressovsky (1944).

4. Autonomy or Heteronomy of Living Things: Ontological Questions

Both from the side of the philosophy of nature and from that of empirical biology, well-intended attempts have repeatedly been made in recent years to bridge the chasm which exists between the two ways of thinking. In these efforts attention is often deliberately restricted to the question whether an unambiguous "characterization" of living things is possible without the assumption, at the same time, of ontological statements in the metaphysical sense. Or the question is raised how far a special type of conformity to law could be found in the organic world which would allow a special position to the methods of research in this domain of nature. Moreover, the question has often been regarded as decisive whether these special laws of the living are reducible to the laws which hold in the inorganic world. In this way the question arises which stands at the center of many discussions—the question of the "autonomy or heteronomy" of living things. In many of these discussions there is a lack of unambiguous definitions—for instance, concerning the demands which are to be made of a general "characterization" of living things or of the fundamental "peculiarity" of a law.

Von Bertalanffy, especially, has tried, in numerous weighty writings, beginning on the basis of an extensive biological knowledge, to bridge by means of an "organismic view" the antitheses between the various speculative systems in biology. Supported by the guiding principle of assuming "a special form of law peculiar to living things," he tries to avoid all metaphysics and is therefore rejected by most vitalists as a mechanist. In his earlier works we find such formulations as "alongside chemical differentiation, still another factor, a special form factor, plays a part." But such an analysis of an organic event into separate processes, divided in accordance with the morphological and physicochemical points of view but without a metaphysical conceptual realism, is meaningless and for that reason can never lead to heuristically useful investigations. The same holds, for example, for the following alternatives set up by von Bertalanffy (1930/31): organizer action is *either* "chemical," conditioned by materials having a formative action, *or* "dynamic," in so far as in the organizer "stronger formative powers are localized" and "an energy gradient exists between it and the environment," in which case it "forces its dynamic" upon its environment, and so "material differentiations are created by the formative movements." If from such statements all vitalistic, executive-causal, and anthropomorphic elements are removed, then these formulations cease to assert anything. Like von Bertalanffy, Ungerer (e.g., 1942) does not wish to be classed as a vitalist but nevertheless uses vitalistically colored statements as guiding principles of a "total or organismic outlook." A noteworthy part is played in the writings of these and other authors by the distinction between "qualitative" and "quantitative" processes, these being opposed to one another as alternatives in the description of formative processes. A distinction is also often made between qualitative and quantitative "causal laws" in biology. But the concepts "quality" and "quantity" have a distinguishing value only for our perceptions of things and not for the scientific reproduction of experiences. It is true that to aid the imagination in predominantly descriptive discussions, especially of a morphological kind, we are accustomed to use the terminology of our "qualitative" ideas; yet we could also reproduce all these facts in a language which, although more prolix, uses only "quantitative" expressions. For a principal distinction or characterization of particular processes or laws, the alternative of "qualitative" or "quantitative" is in any case not useful.

It is interesting that in the course of the development of his organismic view von Bertalanffy has gradually laid aside the vitalistic elements of his theory, so that his characterization of living things today coincides with the view of the organism as a complex open system, which is customary elsewhere in biology. von Bertalanffy has made valuable contributions to the theory of open systems (e.g., 1949, 1950). Ungerer (1942), Alverdes (1939), and some other theoreticians often define the concept of "wholeness," to which they give a central place in their systems, in such a way that it quite, or almost quite, coincides with the biological concept of system. Their demand for a point of view which considers the whole organism is perhaps due to the fact that the unfortunately unavoidable and far-reaching specialization of modern biologists has often led to simplifications and one-sidedness in the representation of biological facts which do not do justice to the system character of living things. This demand is certainly justified. But, unfortunately, behind its residue or kernel there lurks a parabiological view, especially when a general characterization of living things is attempted from a "wholeness" point of view. Living things can certainly be distinguished by a series of peculiarities from nonliving things. This characterization becomes all the clearer the more special the statements become. For a completely satisfactory characterization the whole complex of biological statements must, of course, be invoked. The more we try to make this characterization general, the more the statements become lost in indefinite formulas. The parabiological character of such formulations becomes especially clear when it is sought to distinguish living things from nonliving ones by the "introduction" of a single special "force," of a special "principle," or of a special "category." What scientific biology is able to contribute to the characterization of living things will never be able to satisfy the wish of the metaphysician for an "ontological" definition of the living. The assertions of an empirical science are never "last words" or "reality statements" in the sense of ontology. The empirical scientist must decline to reinterpret the theorems of his particular science as ontological ones and thereby to rob them of their proper meaning, as, unfortunately, often happens from the side of the philosophy of nature (cf. Wenzl, 1938). The same holds for the sharp delimitation between living and nonliving or even between animals and plants, again and again demanded by metaphysics, as well as for the special importance which is attributed to the question of the origin of life on the earth. When Troll (1951) devoted much labor and sagacity to deciding whether, "in the view of ontology," the true viruses are living or not, this really only showed that suggestions toward further developments in the empirical sciences were hardly to be obtained from such considerations. Yet these and other boundaries erected in the name of ontology might prove to be unfortunate for this development if we wished to treat them as untouchable obstacles to the methods of research.

The question of a special kind of law for living things, of "laws special to biology," is particularly dear to the heart of many theoreticians. Some satisfy themselves with the general requirement of such laws or with the general reference to the "character of wholeness" of such laws. We have pointed out in our discussions about the various aspects of biology that it is not possible, from the standpoint of methodological criticism, to call some of its statements more and others less "biological." Nevertheless, some biologists are inclined to consider morphological and ecological statements or statements belonging to the study of behavior "biological" statements, but purely physiological or biochemical statements they regard as "nonbiological." Behind such distinctions is concealed the above rejected assumption of "living" and "nonliving" parts of organisms, the assumption of a special specific "activity" in organisms or a mechanistically or vitalistically oriented idea of wholeness. Where concrete examples for "specifically biological laws" are brought forward, we find mentioned, for example, the following: the allometry law, certain laws relating to metabolism, Child's gradient theory, Mendel's rules, certain phylogenetic laws which can be derived from the evaluation of series of fossil forms, and so forth. In all these cases we are dealing with formulations having the type of genuine empirical scientific statements. What distinguishes them from other biological statements, although of course not sharply, is the fact that they are blanket statements, statements about a very complex occurrence, the initial and final states of which they connect causally by giving a rule. It is clear to everyone that such statements represent the subsumption of a great number of single processes which take place within the system and in reciprocal action with the environment and are causally connected in the most complicated manner with one another and with the environmental constellations. Many of these laws are of the classical statistical type, in which the individual processes, about which they say something collectively, certainly take place on the macrophysical level and according to classical causality. Such blanket statements are not, however, at all confined to biology but occur in every branch of natural science, as in the description of geological processes. The fact that they seem to be especially common in biology depends, indeed, only on the high degree of complication of the processes about which something is to be said. These assertions only preserve their empirical scientific character if we define the concepts occurring in them in a manner which is empirically testable. Statements about "Gestalten" thus only do so if physical structures are intended. But as soon as "form" is viewed in the conceptually realistic sense of idealistic morphology, the statements lose this character. But the same holds also for "form" or "Gestalt" in the statements of inorganic natural sciences and is not therefore the special peculiarity of "specific biological laws."

It is a further question how far blanket statements of this kind can be analyzed into statements about all single processes which take place in the total event, whether they can be reduced to these simpler statements without remainder, or, as is often said, how far the total event can be satisfactorily "explained" in terms of the single processes. It is certainly the case that for none of the examples of biological laws mentioned above has this analysis been carried out completely, nor does the prospect of such an analysis exist. Even the wish for such a *complete* analysis scarcely exists, for it would not be likely to yield any noteworthy results. Moreover, it would occur to no one to demand of a statement expressing laws of geological processes information about the individual fate of every grain of sand or molecule contributing to the composition of the rocks. It seems much more essential that some important individual processes within such a biological total event should be subject to analysis by the most diverse methods and that the laws thus obtained should not stand in contradiction to the blanket statement about the total event. Where such contradictions have seemed to result, they have always led to the discovery of new, hitherto unknown connections or constellations of conditions. But contradictions *in principle* between blanket statements in biology regarded as "specifically biological" and the statements about the single processes felt perhaps to be "less biological" have nowhere resulted. On this circumstance the scientific picture of the world, which is, on the whole, free from gaps and from contradiction, rests. Naturally, this is not to say that a direct reduction of biological blanket statements to some simple laws or "principles" of inorganic nature can be carried out. Such a requirement fails to recognize the high degree of complication of living nature and arises chiefly from the metaphysical requirement of certain philosophical tendencies. The requirement of an answer in principle to the alternative "autonomy or heteronomy of organic life" is thoroughly metaphysical and cannot be made within empirical science.

It is interesting to note what a philosopher who has a sufficient book knowledge of biology, for example, Ballauff (1943), says about this question. It is striking to an empiricist that in the discussion of biological matters (e.g., the "system problem" or the "Gestalt problem") the most diverse theories are brought forward as co-ordinate, equivalent possibilities for thought or are set up in opposition to one another as alternatives, although some of them are formulations that are free from objection from the standpoint of empirical science, while others are pure metaphysical speculations. In this lack of critical examination there arises a chaos of obscurities, contradictions, and absurdities which has very little to do with scientific biology. Ballauff himself feels this and, from the philosophical standpoint, applies the sharpest criticisms against the systems which we call "parabiological," especially against their realism of ideas and concepts and the anthropomorphism of their formulations. He proclaims a "gnoseological crisis" in the present-day philosophy of nature. Even biologists like Ungerer (1942), for example, who wish deliberately to exclude metaphysics from their view, are insufficiently critical in drawing the boundary between methodologically unobjectionable theories and parabiological speculation. Ungerer sympathizes, like some other biologists—for example, M. Hartmann—with the "stratigraphical systems" of the philosophy of nature, as they have been developed as a "theory of levels of being," especially by Nicolai Hartmann. Although the careful formulations of this theory approximate closely the form of empirical scientific theories by the utmost avoidance of metaphysical speculations, yet to the empiricist they seem, as general statements, to be suitable only to serve as tautological statements, not as sources for the derivation of special statements which are empirically testable. It seems to be the case that every attempt to build, on the basis of the progress of the special sciences, by way of an "inductive metaphysics," systems which have more to say than the special sciences is condemned to remain a play of ideas which is not able to help the empirical sciences in their work but may only hinder them.

C. World Picture and Philosophy of Life

The restricted and very doubtful function which speculative theories today exercise in the real working of biology, and in the whole structure of the empirical natural sciences, would scarcely justify the efforts which so many thinkers devote to this field and the many interested readers which publications of this sort find. Between the speculative systems of biology and philosophical, literary, religious, and political systems of thought, we find all kinds of transitions. Among these are found systems of high aesthetic charm and of deep ethical content. Philosophical speculation about nature obviously exercises psychological and social functions which lie within the domain of philosophies of life.

The Significance of Speculation in Biology

Many theoretical biologists, for example, Ungerer, Driesch, etc., especially emphasize the importance of biological theory for a world view and from this derive the demand for a stronger emphasis on speculation in biology. The true basis for the growth of speculation thus lies in a need for a philosophy of life.

By 'world picture' (*Weltbild*) and 'philosophy of life' (*Weltanschauung*) we shall here understand well-defined domains of human spiritual life. By 'world picture' will be meant the total representation of the world which we can form on the basis of all statements of empirical science. It seeks to embrace everything which can be accessible to us on the basis of outer experience. The world pictures which various peoples at various times have formed have been very diverse, according to the way of handling empirical methods and the intellectual elaboration of the material of experience. Today, also, the world pictures of different contemporaries are very diverse in content and extent according to the degree of education in empirical science and the spiritual capacity of the individual. A world picture can, in any case, be objectified in principle in a transtemporal and transindividual manner, in so far as it has outer experience as its basis, as a common gift to all men. The world picture is liable to errors and incompletenesses which can be corrected by outer experience. In their practical behavior for the attainment of external goals men guide themselves according to the current world picture. From this practical application of the world picture there result the applied sciences, technology, medicine, agriculture, and the like.

On the other hand, by 'philosophy of life' is to be understood the totality of all the emotionally tinged strivings and ideas of a man, the totality of his value-judgments, of his ethical and aesthetic maxims. Ideas about the meaning and the tasks of human life, about behavior toward his neighbors, about the meaning of human societies, and therewith about the whole field of ethical, social, religious, and political activities of a man are rooted in his philosophy of life. While a man thus reaches decisions regarding inner and outer goals of his personal and suprapersonal life on the basis of his philosophy of life, he obtains the means for the attainment of outer goals from his knowledge of the world.

It is the conviction of the author of this monograph that *every* philosophy of life rests on a faith, on a decision to trust, which can only be reached from an inner human experience. But on the basis of the world picture of empirical science a philosophy of life can never be built. From the statements of empirical science not a single decision in matters concerning a philosophy of life or valuation can be reached. Biology as an empirical science can therefore never give an answer to those "great questions of life" which move men from within. Naturally, this does not mean that occupying one's self with biology or with another empirical science or the mere contemplation of the empirical world picture is not an experience of high ethical and aesthetic content. Yet the sources of valuation or of emotion flow from the domain of the philosophy of life. The synthesis between philosophy of life and world picture, as a spiritual task of every time and of every man, is therefore never only a matter of empirical science but always a matter of faith. The author of this monograph, who counts himself fortunate in having, as a Catholic Christian, a positive belief, has always felt this consistent synthesis to be the greatest fulfilment of the spiritual life. Nevertheless, he believes that the strong emphasis on speculation in the empirical science, and the various philosophical systems of nature found within modern European spiritual history are symptoms of the weakness of faith and the uncertainty and disunity of our times. Many of these systems clearly bear the stamp of religious substitutes. They seek to fill the substance of faith, hollowed out by the Enlightenment and by liberalism, with pseudo-scientific content. In this way elements foreign to science are imposed upon the empirical sciences, and their methodological foundations are threatened.

This becomes especially clear in the great totalitarian ideologies of our time, which have taken over this tendency as a spiritual inheritance of the nineteenth century. In their efforts to subordinate all the religious and philosophical convictions and feelings of mankind to their program, at the same time denying the latter's foundation in belief and pseudo-religion, they attempt a scientific justification of this program on a basis of philosophy and empirical science. German National Socialism attached great importance to the parabiological "wholeness" theories, because it wished to justify pseudo-scientifically on this basis the totalitarian demands of *Volk* and *Rasse*. Marxism transfers the same mode of argument to "class." Of the biological subdisciplines, genetics, especially, is repeatedly drawn into the conflict of opinions about philosophies of life. While the representatives of exact genetics were called "materialists" by the vitalists, they were branded as "idealists" by the biologists of the school of Lysenko. The philosophical and political ideologues would also like to make their doctrines the foundation and measuring rod for the method of natural science. The author of the present work heard in a programmatic speech of a leading German biologist in the year 1938 these words: "The connection [between "race" and culture] is intuitively grasped by the *Führer*, and it is the one and only task of German science to buttress this intuition scientifically." Similarly, in the condemnation of exact genetics by Lysenko the highest authority was the pronouncement of the party leader, not scientific observation.

IV. Conclusion

In this monograph an attempt has been made to show that biology, in the structure of its statements and in its methods of work, is a purely empirical science. Corresponding to the peculiarity of its object, it has developed special methods and special subdivisions, but these are nevertheless not different in principle from those of other natural sciences. Its assertions, in so far as they exceed the boundaries of pure descriptions, have the form of empirical scientific hypotheses or theories. The explanatory value of its statements rests on foundations similar to those of other natural sciences. The derivative modes of expression or tautological formulations of certain statements which aid picturability are, corresponding to the peculiarity of its object, perhaps more frequent in biology than in other natural sciences but do not detract from its empirical character.

As an unavoidable consequence of its rich development, biology has experienced an especially marked subdivision into special branches, and this carries with it a certain danger of one-sidedness. The synthesis of the results of biology nevertheless goes on throughout consistently and fruitfully and leads to a constant development of the science. There is in biology no "crisis," as has sometimes unjustly been stated. The synthesis of the results of biology with those of the remaining natural sciences has been fruitfully established in many borderline regions and leads to an empirical world picture which is on the whole consistent and unified, if incomplete.

A critical study of foundations and methods which could only be hinted at in this monograph would certainly be very useful in biology. But here biology occupies no special position, because such problems are common to all of the empirical sciences.

Selected Bibliography

ALVERDES, F. 1939. "Biologische Ganzheitsbetrachtung," *Zeitschrift für die gesamte Naturwissenschaft*, Vol. V.

BALLAUFF, T. 1940. "Über das Problem der autonomen Entwicklung im organischen Seinsbereich," *Blätter für deutsche Philosophie*, Vol. XIV.

———. 1943. "Die gegenwärtige Lage der Problematik des organischen Seins," *ibid.*, Vol. XVII.

———. 1949. *Das Problem des Lebendigen*. Bonn: Humboldt-Verlag.

BERTALANFFY, L. VON. 1932, 1942. *Theoretische Biologie*,Vols. I and II. Berlin: Verlag Borntraeger.

———. 1949. *Das biologische Weltbild*. Bern: A. Francke A.G. (English trans., 1952, *Problems of Life* [New York: John Wiley & Sons; London: Watts & Co.]).

———. 1950. "An Outline of General System Theory," *British Journal for the Philosophy of Science*, Vol. I.

BÜNNING, E. 1935. "Sind die Organismen mikrophysikalische Systeme?" *Erkenntnis*, Vol. V.

———. 1949. *Theoretische Grundfragen der Physiologie*. 2d ed. Stuttgart: Piscator-Verlag.

Selected Bibliography

DRIESCH, H. 1928. *Philosophie des Organischen.* 4th ed. Leipzig: Quelle & Meyer.

———. 1944. *Biologische Probleme höherer Ordnung.* 2d ed. Leipzig: Barth.

FRANK, P. 1932. *Das Kausalgesetz und seine Grenzen.* Vienna: Springer.

———. 1950. *Modern Science and Its Philosophy.* Cambridge, Mass.: Harvard University Press.

HARTMANN, M. 1948. *Die philosophischen Grundlagen der Naturwissenschaften.* Jena: Fischer.

HEMPEL, C. G. 1952. *Fundamentals of Concept Formation in Empirical Science,* in *International Encyclopedia of Unified Science,* Vol. II, No. 7. Chicago: University of Chicago Press.

LENZEN, V. F. 1938. *Procedures of Empirical Science,* in *International Encyclopedia of Unified Science,* Vol. I, No. 5. Chicago: University of Chicago Press.

MAINX, F. 1932. *Die Sexualität als Problem der Genetik.* Jena: Fischer.

MITTASCH, A. 1938. *Katalyse und Determinismus.* Berlin: Springer.

MÖGLICH, F.; ROMPE, R.; and TIMOFÉEFF-RESSOVSKY, N. W. 1944. Über die Indeterminiertheit und die Verstärkererscheinungen in der Biologie," *Naturwissenschaften,* Vol. XXXII.

SCHRÖDINGER, E. 1944. *What Is Life?* New York: Cambridge University Press.

SPEMANN, H. 1936. *Experimentelle Beiträge zu einer Theorie der Entwicklung.* Berlin: Springer.

TROLL, W. 1948. "Urbild und Ursache in der Biologie," *Sitzungsberichte der Heidelberger Akademie der Wissenschaften,* Vol. VI.

———. 1951. *Das Virusproblem in ontologischer Sicht.* Wiesbaden: Steiner.

UNGERER, E. 1942. "Die Erkenntnisgrundlagen der Biologie," in *Handbuch der Biologie,* Vol. I. Potsdam: Athenaion.

WENZL, A. 1938. *Metaphysik der Biologie von heute.* Leipzig: Meiner.

WINTERSTEIN, H. 1928. *Kausalität und Vitalismus vom Standpunkt der Denkökonomie.* 2d ed. Berlin: Springer.

———. 1938. "Der mikrophysikalische Vitalismus," *Erkenntnis,* Vol. VII.

WOLTERECK, R. 1931. "Vererbung und Erbänderung," in H. DRIESCH and R. WOLTERECK (eds.), *Das Lebensproblem.* Leipzig: Quelle & Meyer.

WOODGER, J. H. 1939. *The Technique of Theory Construction,* in *International Encyclopedia of Unified Science,* Vol. II, No. 5. Chicago: University of Chicago Press.

ZIMMERMANN, W. 1938. "Strenge Objekt/Subjekt-Scheidung als Voraussetzung wissenschaftlicher Biologie," *Erkenntnis,* Vol. VII.

———. 1948. *Grundfragen der Evolution.* Frankfurt an der Oder: Klostermann.